Information Systems Foundations:

THEORY BUILDING IN INFORMATION SYSTEMS

Information Systems Foundations:

THEORY BUILDING IN INFORMATION SYSTEMS

Dennis N. Hart and Shirley D. Gregor (Editors)

E PRESS

Published by ANU E Press
The Australian National University
Canberra ACT 0200, Australia
Email: anuepress@anu.edu.au
This title is also available online at: http://epress.anu.edu.au

---

National Library of Australia Cataloguing-in-Publication entry

Author: Information Systems Foundations ('Theory Building in Information Systems') Workshop (2010 : Canberra, A.C.T.)

Title: Information systems foundations : Theory Building in Information Systems/ edited by Dennis N. Hart and Shirley D. Gregor.

ISBN: 9781921862939 (pbk.) 9781921862946 (ebook)

Notes: Includes bibliographical references.

Subjects: Management information systems--Congresses.
Information resources management--Congresses.

Other Authors/Contributors:
Gregor, Shirley Diane.
Hart, Dennis Neil.

Dewey Number: 658.4038

Cover design by ANU E Press

Cover illustration by Michael Gregor

Contents

Part Three: The Big Picture

Contributors

Workshop Chair

Shirley D. Gregor, The Australian National University

Program Chairs

Shirley D. Gregor, The Australian National University
Dennis Hart, The Australian National University
Jan Recker, Queensland University of Technology

Program Committee

Andrea Carugati, Aarhus University, Denmark
Patrick Chau, University of Hong Kong, Hong Kong
Walter Fernandez, The Australian National University, Australia
Guy Gable, Queensland University of Technology, Australia
Sigi Goode, The Australian National University, Australia
Peter Green, University of Queensland, Australia
Dirk Hovorka, Bond University, Australia
James Jiang, The Australian National University, Australia
Robert Johnston, University College Dublin, Ireland
Aleck Lin, The Australian National University, Australia
M. Lynne Markus, Bentley University, USA
Nigel Martin, The Australian National University, Australia
Craig McDonald, University of Canberra, Australia
Arun Rai, Georgia State University, USA
Michael Rosemann, Queensland University of Technology, Australia
Tracy Ann Sykes, The Australian National University, Australia
Eric van Heck, Erasmus University, Netherlands
John Venable, Curtin University, Australia
Iris Vessey, University of Queensland, Australia
Xiaofang Zhou, University of Queensland, Australia

Preface

This volume contains the papers presented at the Information Systems Foundations Workshop, 30 September – 1 October 2010. This workshop was the fifth in The Australian National University (ANU) series of biennial workshops that was originally inspired by one held in 1999 by Kit Dampney at Macquarie University, and that focuses on the theoretical foundations of the discipline of information systems (IS).

The theme of the 2010 workshop was 'Theory Building in Information Systems' and it once again allowed researchers and practitioners in the field of information systems to come together to discuss some of the fundamental issues relating to our discipline.

Information systems is still a relatively young field of study that, perhaps uniquely, is a peculiar juxtaposition of the technological, in the form of computing and communication technology, and the non-technological, in the form of the humans and organisations that design, implement and use systems built with that technology. It is, indeed, this juxtaposition of artefacts and phenomena from what are often termed the 'hard' and the 'soft' ends of the spectrum that makes the development of useful and robust theory in the field such a challenge. The sciences that underlie and deal with technologically oriented fields such as computing and related areas generally result in theories that fit within the 'covering law' model—that is, are assumed and believed to have universal applicability and explanatory and predictive power. And, typically, such theories exhibit a deterministic character as well. By contrast, theories in the human sciences are generally much more circumscribed in the phenomena with which they deal and the domains in which they are applicable, and are much more conditional, contextual, tentative and open to exceptions. Trying, therefore, to build successful theory in a discipline like information systems in which phenomena from both the 'hard' and the 'soft' coexist is a bit like trying to mix together immiscible fluids such as oil and water: it doesn't matter how much you stir it all up, what you wind up with is still just a conglomeration of little blobs of oil and other separate little blobs of water! It was, therefore, with this kind of problem in mind that the 2010 workshop was held.

Typically the information systems foundations workshops give authors an opportunity to present papers and get feedback on ideas that might be regarded as too new or risky for publication in conventional outlets. There have been some good outcomes from this approach, with revised papers going on to find a wider audience in mainstream journals. As the workshop is deliberately kept small, and there is only one stream of papers, all paper presentations are typically attended by all participants, which often leads to ongoing and vigorous discussion.

The papers presented here were accepted after a double-blind review process and we thank our program committee and reviewers for their assistance. We also acknowledge and thank the sponsors of the workshop: the National Centre for Information Systems Research (NCISR), the Australian Research Council (ARC) Enterprise Information Infrastructure (EII) network and the School of Accounting and Business Information Systems at ANU.

Finally, we would like to thank the keynote speakers at the workshop—Mike Morris, Viswanath Venkatesh and Ron Weber—whose presence, expertise and participation added greatly to the value of the event for all concerned. All in all, therefore, the workshop provided a stimulating and productive as well as an enjoyable couple of days for both the authors and the attendees, and we hope that the papers that form this volume will provide similar stimulation, provoke similar productive outcomes and perhaps provide some enjoyable reading as well, for a wider audience than those who were able to attend the workshop itself.

Shirley Gregor

Dennis Hart

The Papers

The 11 papers in this book are organised into three sections entitled 'Fundamental Issues', 'Theories and Theorising in Practice' and 'The Big Picture', reflecting the wide range of topics relating to theories and theory building that were addressed in the 2010 workshop.

The fundamental issues section contains three papers, each of which puts forward a particular perspective on how theories in information systems should be structured or built. The first, by Weber, offers a view of what theory is, or should be, through a set of criteria for evaluating theory quality as well as a detailed example of how those criteria can be applied. Following this, Tate and Evermann identify a number of obstacles that they argue can, and in many instances do, seriously hinder the development of good theory in the field. Importantly, however, beyond just identifying these obstacles they also offer suggestions as to how they might be avoided or overcome. The third and final paper in the fundamental issues section is by Hovorka and Gregor, who tackle the longstanding philosophical conundrum of causality, though in the quite specific context of design science and its application in information systems. The result of their effort is a framework that they propose can be used to identify the type of causal analysis suitable for different types of theorising in designing new, and especially novel, artefacts of an information systems type.

The 'Theories and Theorising in Practice' section, which follows that on fundamental issues, constitutes the bulk of the book. It contains seven papers that discuss specific theories in information systems or the process of building them. The first, by Fidock and Carroll, is concerned with theories that deal with the entire life cycle of an information system. In it, the authors outline and critique the most well known of such theories before proposing their own, based on the 'Model of Technology Appropriation', which is aimed at overcoming the shortcomings of the other theories they consider. Next, Raza and Standing focus on a particular aspect of the life cycle—namely, that of coping with multiple stakeholders and their different interests and perspectives during the system development process. They base their analysis on critical systems thinking (CST) and propose a process of what they call 'phase-stakeholder-identification' as a tool for use by project managers engaged in information systems development in a multi-stakeholder environment, as most are.

The third paper in the 'Theories and Theorising in Practice' section is by Hoehle and Huff. Unlike the previous two papers, which aim for validity and relevance across all information systems application areas, these authors consider a particular theory (task-technology fit, or TTF, theory) in a particular context (electronic banking). More specifically, they analyse in some detail the central concept of TTF—that of 'fit'—and then go on to use their analysis to devise and test a measurement instrument for determining the degree of 'fit' between various electronic banking tasks and the channel(s) over which they are conducted, aiming, all the while, at advancing TTF theory through their efforts. The next paper, by Koeglreiter, Smith and Torlina, changes tack again because its authors are interested in the *process* of research and theory development in general rather than the content of any specific theory, as were Hoehle and Huff. In particular, Koeglreiter and her co-authors describe how, in their own research, they have developed an integrated method, which they call 'structured-case with action interventions', that melds together the action research and case-study methods, thereby taking advantage of the strengths as well as avoiding the weaknesses of both.

In the next paper, Hasan and Banna present an argument for making use of 'activity' as a unit of analysis in information systems theory. Their case is based on the 'activity theory' of the Russian psychologist Vygotsky and his successors, of which they give a brief overview. They then follow this with an example of a research project in which they were involved and which they reinterpret using 'activity' as the unit of analysis to illustrate the benefits of doing so. The final two papers in the 'Theories and Theorising in Practice' section both report on research in progress. The first, by Mola, Rossignoli, Fernandez and Carugati, describes a continuing study of a group of agricultural cooperatives in Italy that is undertaking an extensive modernisation process, including their information and communication technology capabilities. The aim of the project, ultimately, is to achieve a better theoretical understanding of modernisation efforts of this type in the agricultural sector. The second 'research in progress' paper and the final paper in the section, by Dod and Sharma, is concerned with business analytics. After introducing the background to the research, it outlines the theory-building effort to be pursued in their future work.

The final section of the book, entitled 'The Big Picture', contains just one paper, by McDonald. This invited paper takes a critical, high-level and broad-ranging look at what theory, and 'grand theory' in particular, is from what the author terms an 'informatics' perspective based on the ideas of the philosopher Karl Popper. It argues and concludes that theories are, in fact, systematic patterns that, being themselves information constructs, should naturally be 'susceptible to examination and systems building by the IS discipline' but that are currently rather poorly served in this sense. Though this might be the case, it remains to be seen exactly how such examination and building might be effected at the level of theory (and especially 'grand' theory), and what the benefits to be expected from such efforts might be. It will be interesting to see.

Dennis Hart

Shirley Gregor

Part One: Fundamental Issues

1. Theory Building in the Information Systems Discipline: Some critical reflections

Ron Weber
Monash University

Abstract

This chapter articulates criteria for evaluating the quality of a theory. It also shows how the criteria can be used to pinpoint the strengths and weaknesses of a theory that, based upon citation evidence, appears to have had a significant impact on other researchers within the information systems discipline. Aside from their evaluation purposes, the criteria are also intended to inform researchers who are seeking to build high-quality theory to account for some type of phenomena within the information systems domain.

> Nothing is as practical as a good theory. (Lewin, 1945, p. 129)

Introduction

For many researchers, the development of theory within their discipline is the central goal—the 'jewel in the crown'—of their research endeavours (for example, Eisenhardt, 1989). By articulating high-quality theory, they believe they are more likely to enhance their own knowledge of, other scholars' knowledge of and practitioners' capabilities to operate effectively and efficiently in their domain of interest.

In spite of the importance ascribed to theory by many researchers, the development of theory has been a relatively neglected feature of research within the information systems (IS) discipline. As a result, in the late 1990s and early 2000s, several editors of major journals appealed for more theoretical contributions to be made to the discipline (for example, Zmud, 1998). Moreover, only recently have scholars within the discipline begun to focus on how *high-quality* theory can be developed (for example, Gregor, 2006). Nonetheless, in many respects, the process that should be used to develop high-quality theory remains an arcane affair.

In this chapter, I propose a set of normative criteria that can be used to evaluate the quality of a theory. I show how these criteria can be used to pinpoint the strengths and weaknesses of a theory that has been proposed within the information systems discipline—one that according to citation evidence has had a significant impact on many researchers within the discipline. In this way, I seek to illustrate the usefulness of the criteria. As a related matter, I seek to show also that the criteria provide useful guidelines for *developing* theory.

The structure of the chapter is as follows. First, I briefly define some ontological constructs that enable me to define the meaning I ascribe to the term 'theory' and to articulate the criteria I propose for evaluating the quality of a theory more precisely. Next, I explain the meaning I ascribe to the term 'theory'. I then describe the criteria that I propose for evaluating the quality of a theory. Subsequently, I attempt to show the usefulness of these criteria by applying them to the evaluation of an important, extant information systems theory. Finally, I provide some brief reflections and conclusions.

Some Basic Ontology

To provide the basis for my analyses in the sections that follow, this section provides a brief (and somewhat informal) explanation of some fundamental ontological constructs. These constructs are derived from a formal ontological theory proposed by Bunge (1977, 1979).

- *Thing*: The world is made of things. Things can be *substantial* or *concrete* (for example, people or buildings); alternatively, they can be *conceptual* (for example, sets and functions). In this chapter, my focus is primarily on *concrete* things.
- *Composite thing*: Some things are made up of other things (for example, a team, which is a composite thing, is made of team members, which are its components).
- *Property*: All concrete things in the world possess properties (there are no formless things). Similarly, all properties in the world attach to some thing (properties do not exist in isolation from things). Properties are *not* things; they are separate ontological constructs that describe different elements in the world.
- *Class*: Things that possess at least one property in common constitute a class of things.
- *Attributes*: We 'know' about properties of things in the world through our *perceptions* of them. These perceptions might be more or less true. The way

in which we perceive a property at a point in time (our representation of it) is called an attribute.

- *Types of attributes*: Various types of attributes exist
  - *intrinsic attributes* represent properties of individuals (for example, the height of a person)
  - *mutual attributes* represent properties of two or more things (for example, the date on which one person was married to another person)
  - *emergent attributes* are attributes of composite things that do not belong to their components but nonetheless are related in some way to attributes of their components (for example, the work productivity of a team has no meaning in terms of each team member, but it is related in some way to the productivity of each team member)
  - *attribute in general*: attributes in general belong to a class of things (for example, all humans possess the attribute called 'height').
  - *attribute in particular*: particular things in a class of things possess attributes that have particular values (for example, the thing called 'John' in the class of things called 'people' possesses the particular attribute 'height is 180 centimetres').
- *State*: A vector of attributes in particular represents a state of a thing (its attributes in general along with their associated values). Some states are deemed *lawful* (they obey natural or human-made laws); others are deemed unlawful.
- *Event*: An event that a thing undergoes is represented by a change from one of its states to another of its states. Some events are deemed *lawful*; others are deemed unlawful. If an event has an unlawful beginning or end state, the event will be unlawful. Some events are unlawful, however, even when their beginning and end states are *lawful*. For instance, 'alive' and 'dead' are lawful states of a human thing. The event represented by the state change from 'alive' to 'dead' is lawful; the event represented by the state change from 'dead' to 'alive' is unlawful.
- *Interaction*: Two things interact when the history of one thing, which is manifested as changes in its states (or attributes in particular), is not independent of the history of the other thing.

In the sections below, I use these constructs to explain the nature of theory and criteria that can be used to evaluate the quality of a theory.

Nature of Theory

Different researchers often ascribe different meanings to the term 'theory'. For instance, Gregor (2006) examines five ways in which the term 'theory' has been used in the literature. I do not agree that the term 'theory' covers all the meanings she canvasses (in my view, those she calls Type-I and Type-V 'theories' are not theories). For this reason, in this section I explain the meaning I ascribe to the term 'theory'.

By *theory*, I mean a particular kind of *model* that is intended to *account* for some subset of *phenomena* in the world. A theory is a *social construction*. It is an artefact built by humans to achieve some purpose. It is a *conceptual* thing rather than a concrete thing.

By *phenomena*, I mean changes in the *attributes in particular* of some things in the world. The subset of phenomena in the world that the theory is intended to cover is called the *focal phenomena*.

By *account*, I mean a theory assists its users to *predict* and/or *explain* its focal phenomena. Some researchers ascribe another purpose to theories—namely, to facilitate human *understanding* of the theory's focal phenomena. I do not see how *explanation* of focal phenomena can occur, however, without first *understanding* the focal phenomena. For this reason, I intend the purpose of explanation to encompass the purpose of understanding.

By *model*, I mean a *representation* of something else (phenomena) in the world. Theories are, however, particular kinds of models (see section four below). All theories are models, but not all models are theories. A model must satisfy particular conditions before I deem it to be a theory (see below).

Framework for Theory Evaluation

In this section, I argue that a theory must be evaluated from two perspectives. The first is the 'parts'—the evaluation must focus on the quality of the individual components that make up the theory. The second is the 'whole'—the evaluation must focus on the quality of the theory considered *in toto*. Both forms of evaluation are important in assessing the quality of a theory. Clearly, it is unlikely that the quality of the whole will be high if the quality of the parts is not high. Nonetheless, high-quality parts are a *necessary* but not *sufficient* condition for a high-quality whole. To the extent a *model* satisfies these criteria, it can be deemed a *theory*.

Parts

A theory has three parts (or components): its constructs, its associations and its boundary. When evaluating a theory, the focus initially should be on the quality of these parts. The following subsections explain the nature of each part. They also describe criteria that can be used to evaluate how well a researcher has articulated each of the parts.

Constructs

A *construct* in a theory represents an *attribute in general* of some *class of things* in the focal domain (as opposed to a *particular attribute* of a *specific* thing). The classes of things to which attributes in general pertain ought to be defined precisely to ensure the *meanings* of each class and the things in each class are clear. Otherwise, the meanings of the attributes in general that attach to the classes of things are unlikely to be clear. Attributes do not float in the ether; they always attach to things. As a first step in clarifying the meaning of an attribute, therefore, the thing to which it attaches needs to be made clear.

Once the meanings of the classes of things that a theory covers are clear, the nature of each attribute in general that pertains to a particular class ought to be defined precisely. Unless the meanings of the attributes in general are clear, the meanings of any associations among the attributes in general cannot be clear. Moreover, developing credible (valid and reliable) empirical indicators of the attributes in general will be difficult (if not impossible).

Associations

An association between two constructs in a theory shows that a *history of an instance* of at least one of the constructs is conditional on a *history of an instance* of the other construct. In other words, at least *one change in the value of an instance of one construct* is somehow related to at least *one change of value in an instance of the other construct.*

If two constructs represent different attributes in general of a *single* class of things, any association between them means the two attributes are lawfully related. In other words, for at least one instance of a thing in the class, a change in the value of at least one of the two attributes is related to a change in the value of the other attribute.

If two constructs represent different attributes in general of two *different* classes of things, any association between them means at least one instance of a thing in one class *interacts* with at least one instance of a thing in the other class. In other words, the histories of the two things are not independent of each other. The nature of the interactions between the two things is manifested in the attributes that are related.

Associations can be specified with varying levels of precision (Dubin, 1978).

- Two constructs are simply shown to be related to each other, but neither the 'sign' nor the 'direction' of the association is shown. In other words, the association does not imply causality, nor does the association indicate whether a positive or negative change in the value of an instance of one of the constructs is associated with a positive or negative change in the value of an instance of the other construct.
- The sign of the association between two constructs is shown, which implies that changes in the values of an instance of one of the constructs are positively or negatively correlated with changes in the values of an instance of the other construct.
- The 'direction' of the association between two constructs is shown, which implies causality (some changes in value of an instance of construct A cause a change in value of an instance of construct B) or at least a time series of value changes (a change in the value of an instance of construct A precedes a change in the value of an instance of construct B).
- A functional association is shown between two constructs. In other words, the amount of change that occurs in the value of an instance of one construct as a result of a change that occurs in the value of an instance of another construct.

To the extent that the nature of the associations among constructs in a theory can be specified more precisely, empirical evaluations of the theory can be done more precisely. Moreover, to the extent that empirical tests of the theory support the existence of the associations, the theory has greater predictive and/or explanatory power.

When one of the associations in a theory shows directionality, the theory is a *process* theory (Markus and Robey, 1988). The reason is that the directionality means a change in the value of one of the constructs in the association *precedes* a change in the value of the other construct in the association.

Boundary

The boundary of a theory circumscribes the state space and event space of things for which the theory is posited to have predictive and/or explanatory power. A first step in specifying the boundary of a theory, therefore, is to be clear about the class or classes of things that the theory covers and the attributes in general of this class or these classes of things that the theory covers. The constructs in the theory can then be defined precisely.

Unless the constructs in a theory are defined precisely, the boundary of a theory cannot be defined precisely. Often theories cover only certain values of each construct. If the constructs are not clear, the attributes of things in the class or classes of things covered by the theory will not be clear. As a result, the values of the attributes covered by the theory cannot be specified clearly.

Even if the constructs in a theory are defined precisely, however, researchers must still specify the *values* of each construct for which the theory is posited to hold. In the context of the theory, these values underpin the 'lawful states' of the theory—the states of things that the theory covers.

Similarly, researchers must specify the *events* that their theory covers. They must consider possible changes in the values of constructs (state changes) and whether the associations they posit exist among the constructs in their theory hold when such changes occur. In some cases, two states that a construct might assume might be covered by the theory, but a transition between these states might *not* be covered by the theory. Those state changes that are covered by the theory constitute the 'lawful events' of the theory.

When articulating the boundary of their theory, researchers must also consider whether their theory covers different *combinations* of construct states and different *combinations* of or *sequences* of construct events. The behaviour of constructs cannot be considered in isolation; instead, the implications for the theory of states of and events in subsets and sets of constructs must be considered.

The boundary of a theory must be specified precisely if precise empirical tests of the theory are to be undertaken. In the absence of a precise specification of a theory's boundary, researchers might test the theory unwittingly in an inappropriate context (one that falls outside its boundary).

Whole

A theory has *emergent* attributes—attributes of the theory *as a whole* rather than attributes of its parts. Many such attributes exist, and researchers often differ in their views of the significance they ascribe to each attribute. Nonetheless, some emergent attributes have widespread acceptance among researchers as being significant in the evaluation of the quality of a theory. The following subsections explain the nature of these attributes and describe criteria that can be used to evaluate the extent to which a theory possesses them.

Importance

The importance of a theory is often assessed via judgments made about the importance of its focal phenomena. Usually, there is little point to having a theory with rigorously specified constructs, associations and boundary conditions if it addresses uninteresting phenomena. The focal phenomena might be deemed important from the viewpoint of practice (improving the effectiveness and efficiency of some entity's activities). They might also be deemed important from the viewpoint of research. Potentially, enhanced understanding of the focal phenomena will provide key insights that will enable theoretical or empirical progress to be made on some problem within a discipline.

Ex ante, it might be difficult to judge the importance of a theory. At the outset, its potential impact on researchers and practitioners might be difficult to assess. Moreover, sometimes a theory provides insights that were not anticipated when it was first articulated. Such insights arise only when researchers engage with the theory and use it as the basis for their empirical work.

Ex post, however, various metrics are available to assess the importance of a theory. For example, the extent to which a theory is cited by researchers provides an indicator of its impact on their work and thus its likely importance to them. Similarly, whether a theory is cited in practitioner publications or underpins consulting work provides an indicator of its importance.

Citation evidence must be treated cautiously when it is used as a proxy for the importance of a theory. Some theories are appealing to researchers because they are relatively simple to test empirically. They are perceived as an easy route to journal publications. Whether they provide deep insights into the phenomena they cover, however, is another matter.

Novelty

The extent to which a theory is novel appears to be an important factor in determining: a) the value ascribed to it by researchers; and b) the likelihood that papers describing the theory will be accepted for publication in major journals (for example, Mone and McKinley, 1993). In short, judgments about a theory's novelty and judgments about its contributions to knowledge appear to be closely related. Moreover, the *importance* of a theory and its deemed novelty appear to be closely related. Nonetheless, some theories might be considered novel but cover phenomena that are regarded as unimportant.

Weber (2003) describes a number of ways in which a theory might make novel contributions to a discipline. For instance, a theory's focal phenomena might not have been covered by prior theories, or the focal phenomena might be framed or conceived in a different way. In essence, the theory being proposed in these cases is a new theory. Alternatively, an existing theory might be modified by adding and/or deleting constructs and associations, defining constructs and associations more precisely or specifying the boundary of the theory more precisely.

A theory will also be deemed novel (perhaps after some time has elapsed) to the extent it changes the *paradigms* used by researchers to investigate phenomena within their discipline (Kuhn, 1996). It will command the attention of researchers if it provides a way of resolving 'anomalies' within their discipline—that is, empirical observations of phenomena that existing theories are unable to explain or predict. It will also command the attention of researchers if it enables them to 'see' or conceive of new and interesting phenomena (phenomena that previously escaped their attention) or re-conceptualise existing phenomena in new and interesting ways. Such theories break the cycle of 'normal science' within a discipline and set a new path for the discipline to follow.

The *quality of the rhetoric* used by researchers to describe their theories also appears to be an important factor in determining the extent to which their theories are deemed novel (Locke and Golden-Biddle, 1997). Because science is a social phenomenon, researchers have to convince their colleagues that their work has value. In this light, the arguments a researcher uses to expound their theory's novelty must be crafted carefully; otherwise, the theory's contribution to knowledge might be overlooked.

After analysing 82 papers published in the *Academy of Management Journal* and *Administrative Science Quarterly* (two high-quality, high-impact journals) between January 1976 and September 1996, Locke and Golden-Biddle (1997) concluded that researchers who had successfully demonstrated the novelty or contribution of their research used two rhetorical strategies. First, they 'legitimise' their work by 'constructing intertextual coherence'. They 're-present and organise existing knowledge so as to configure a context for contribution' (p. 1029). Second, they 'subvert' or 'problematise' the existing literature. They do so to show that opportunities exist for contributions to knowledge. One way in which the novelty of a theory can be assessed *ex ante*, therefore, is to evaluate how well its proponents enact Locke and Golden-Biddle's two strategies.

Parsimony

A theory is parsimonious when it achieves good levels of predictive and explanatory power in relation to its focal phenomena using a small number of constructs, associations and boundary conditions. What constitutes a 'small number' is in the eyes of the beholder. Nonetheless, Miller's (1956) classic paper on the 'magical number seven, plus or minus two' suggests some guidelines. Humans appear able to manipulate about seven 'chunks' of information in short-term memory. In this light, one might predict that researchers would deem a theory to be parsimonious if it has no more than about seven constructs, seven associations and seven boundary conditions (and perhaps the desired number in each case is less than seven).

In building a theory, researchers are often tempted to include *more* constructs, *more* associations and *more* boundary conditions in an attempt to capture the 'richness' of the phenomena that they are seeking to predict or explain (and my experience is that the inclusion of more constructs, associations and boundary conditions is often a frequent request made by the reviewers of journal papers!).

Parsimony dictates, however, that some constructs, associations and boundary conditions must be *omitted* from a theory. In choosing constructs to omit, those whose instances have little variation in their values (states) are likely candidates. In choosing associations to omit, those where few instances of constructs are related are likely candidates. In choosing boundary conditions to omit, those where only a small number of states and events (pertaining to a construct instance) fall outside the boundary condition are likely candidates to omit.

Often, a trade-off must be made between parsimony and a theory's predictive and explanatory power. As the number of constructs, associations and boundary conditions in a theory increases, the theory might be better able to predict and explain the focal phenomena. At some point, however, users of the theory will deem it to be too complex. The goal is to achieve high levels of prediction and explanation with a small number of theoretical components (Occam's razor).

Level

Some theories cover a very narrow, constrained set of phenomena. They are often called 'micro-level' theories. On the one hand, a micro-level theory's constructs, associations and boundaries might be defined precisely. Moreover, its explanatory and predictive powers might be high in relation to the phenomena it covers. Because of the limited range of phenomena it covers, however, it runs the risk that it will be deemed uninteresting.

Some theories cover a broad range of phenomena. They are often called 'macro-level' theories. In some ways, a macro-level theory might be compelling because of the overall insights it provides into many phenomena. Often, however, its constructs and associations are defined imprecisely. Moreover, its explanatory and predictive powers in relation to the more specific phenomena that are a researcher's focus are limited. It runs the risk that it will become discredited because it ends up being a 'theory of everything' in a discipline.

Merton (1957) argues the primary theories used by a discipline ought to be 'middle-range' (or 'meso-level') theories. On the one hand, such theories avoid 'narrow empiricism'. On the other hand, they avoid being so general in their coverage that it is difficult, if not impossible, to test them empirically. Meso-level theories often have value because they link the micro-level world and macro-level world in a discipline.

In spite of the wide acceptance of Merton's idea within many disciplines, the precise meaning of 'middle-range theories' remains problematic (Boudon, 1991). Whether a theory is at an appropriate level is a matter of judgment. Moreover, a level that is too high or too low in one discipline might be an appropriate level in another discipline. Nonetheless, in the context of their discipline, researchers make judgments about whether a theory is at an appropriate level—whether it is too specific or too broad to be interesting and/or useful.

Falsifiability

Most, if not all, theories cannot be *proven* via empirical tests, because it is impossible to test the theory under: a) *all* combinations of values that its *constructs* might assume; and b) *all* combinations of values that its *boundary conditions* might assume. Instead, *support* for a theory grows when its powers of prediction and explanation remain robust across different tests of the theory. If the theory has been articulated clearly, these tests can be designed strategically. They can be used to examine conditions that researchers believe are most likely to lead to the theory being *falsified* (failing the empirical test) rather than supported (Doty and Glick, 1994; Popper, 1990).

To be capable of falsifying a theory, researchers must be capable of generating sufficiently precise predictions about the focal phenomena so they can undertake reasonably exact empirical tests of the theory. If the predictions they are able to generate are so vague that the status of empirical tests they undertake always remains problematic or alternatively the empirical outcomes can always be 'finessed' (explained) using the theory, the value of the theory is undermined.

Using the Evaluation Framework: An example

To show how the evaluation framework I have proposed above can be used, I examine in the subsections below the paper by Griffith et al. (2003), which examines 'the dynamics of knowledge development and transfer in more and less virtual teams' (p. 265). This paper is one of several published in a special issue of the *MIS Quarterly* on the topic of 'Redefining the Organizational Roles of Information Technology in the Information Age'. The stated purpose of the special issue was to 'stimulate *significant and innovative theoretical thought* in response to the dramatic changes that had occurred in the 1990s regarding information technology and the transformational ways in which information technology was being applied to enable new forms of organisations and markets' (Zmud, 2003, p. 195; emphasis in original).

In spite of the special issue's focus on 'significant and innovative theoretical thought', it is unclear whether Griffith et al. (2003) are seeking to articulate a *theory* or a *model* in their paper (recall, I argue above that not all models are theories and that the latter are models that possess specific attributes). On the one hand, they state their paper 'advances theory' (p. 265). Moreover, they articulate a number of propositions, which suggests that their focus is theory building. On the other hand, they present a 'stylised *model*' (my emphasis) of how 'individual and social knowledge...transfers among individuals...becomes available to the members of the team' (pp. 268–9). Moreover, they use the term 'model' frequently throughout their paper. Nonetheless, the legend they give for their Table 1 of their paper (p. 281) is 'Operationalisation of Constructs to Test the *Theoretical Model*' (my emphasis), which suggests their model is indeed meant to be a theory. In any event, for the sake of illustrating how the evaluation framework might be used, I have assumed that Griffith et al. are seeking to present a *theory* of virtualness and knowledge in teams.

Parts

In this subsection, I evaluate the constructs, associations and boundary that Griffith et al. (2003) employ in their theory. Specifically, I focus on how rigorously each is specified in their paper. In this regard, while the evaluation framework I have proposed above pinpoints those parts of Griffith et al.'s theory that need to be assessed, readers of their theory still need to make judgments about how rigorously each part has been expressed (and accordingly my evaluation below reflects my own judgments). Even where judgments about rigour differ, however, the evaluation framework provides a way for researchers to structure their discourse about the quality of a theory's components.

Constructs

Griffith et al. present their constructs at four places in their paper. First, they are shown in Figure 2 (their 'stylised model') of their paper (p. 269). Second, they can be gleaned from the 19 propositions they state in their paper (pp. 271–8). Third, in Table 1 of their paper (p. 281), they state that they 'catalogue the constructs and assessments necessary to test our model' (p. 280). Fourth, they discuss specific constructs at various places in the text of their paper.

A first problem with Griffith et al.'s articulation of their constructs is that inconsistencies exist among those they show in Figure 2 of their paper, those embedded within their propositions and those they list in Table 1 of their paper.

In this regard, Figure 2 of their paper appears to show 17 constructs they employ in their theory. In my reading of their propositions, however, I can identify 29 different constructs that they employ in their theory (see my Table 1.1 below). Yet Table 1 of their paper shows only 15 constructs that must be subject to 'assessments'.

Table 1.1 Constructs in Griffith et al.'s Theory of Virtualness and Knowledge in Teams

No.	Thing	Attribute
1	Team	Level of virtualness
2	Team	Level of transformation of implicit knowledge into explicit knowledge
3	Team	Level of access to extant explicit knowledge
4	Team	Level of proactive effort made to verbalise rules, terminology and descriptions
5	Team member	Level of tacit knowledge acquired from collocated sources transferred to team
6	Team member	Level of tacit knowledge acquired from team-mates
7	Team	Level of difficulty in forming collective knowledge
8	Team	Level of experienced richness of communication
9	Team	Level of collective knowledge
10	Team	Level of collective knowledge accessible via technological tools
11	Team	Likelihood of enacting an independent approach to tasks
12	Team	Level of shared understanding of tasks
13	Team	Level of access to and appropriation of tools and structures that support highly interdependent work
14	Team	Level of interdependence of work
15	Team	Level of shared knowledge
16	Team member*	Level of transition of potential team knowledge to usable knowledge
17	Team member	Level of absorptive capacity
18	Team member	Level of social interaction limited by virtual work undertaken
19	Team	Level of transition of potential knowledge to usable knowledge
20	Team member	Level of connections to relevant communities of practice
21	Team	Level of access to communities of practice
22	Team	Level of tacit knowledge from members' links to communities of practice disseminated within team
23	Team	Level of transfer of potential team knowledge to usable team knowledge
24	Team	Level of transactional memory
25	Team	Level of transactional memory development
26	Team	Level of virtual work
27	Team	Extent to which technologies of organisational systems are used to support transactional memory development
28	Team	Level of synergy
29	Team	Extent of match between team task and technology use

* indicates it is unclear whether the attribute belongs to 'team' or 'team member'.

Prima facie, some of these inconsistencies appear to represent only *naming* inconsistencies. For instance, Figure 2 of Griffith et al.'s paper shows a construct called 'Individualised Knowledge: Implicit', which is cross-referenced to Proposition 2 (P2) in their paper. Based on the label given to this construct, one might expect it refers to the *level* of implicit knowledge that a member of a virtual team possesses. The focus of Proposition 2, however, is on the *extent to which implicit knowledge can be transferred to explicit knowledge*. These are *not* the same constructs, even though the words 'implicit knowledge' are used in both. Furthermore, the 'assessment' (operationalisation) of the 'Individual Knowledge Types: Implicit' construct in Table 1 of Griffith et al.'s paper does not pertain to the extent to which implicit knowledge can be *transferred* to explicit knowledge (the construct used in P2). Rather, it refers to the 'extent to which individuals rely on...knowledge which could be codified but has been made automatic by practice' (p. 281).

A similar problem exists with other constructs—that is, the meaning that prima facie might be assigned to a construct shown in Figure 2 of Griffith et al.'s paper does not match the construct employed in their propositions. In addition, the construct used in their propositions does not match the construct in Table 1 of their paper.

A second problem with Griffith et al.'s articulation of their constructs (which is to some extent a corollary of the first problem) is that some are defined rigorously (for example, the level of team virtualness and 'individual knowledge types'), but others are not. Moreover, the meaning of some constructs has to be elicited from the text the authors use to articulate and support their propositions. Sometimes the meaning of these constructs is clear; sometimes it is not.

For instance, a construct Griffith et al. use in their Proposition 4b is 'Level of Collective Knowledge Accessible Via Technological Tools'. Earlier in their paper (p. 273), they define collective knowledge reasonably precisely as 'explicit knowledge that has been internalised by the team members'. What they mean by 'technological tools', however, is discussed only somewhat obliquely. Nonetheless, it is not clear which of the following meanings they ascribe to the construct they use in Proposition 4b: a) the *nature* of the collective knowledge formed by more virtual teams means that this knowledge is easier to access via 'technological tools'; or b) more virtual teams have more access to or greater facility with 'technological tools' and thus find it easier to access collective knowledge; or c) both meanings apply to the construct. If the theory is to be rigorously operationalised (tested), the meaning of the construct must be clarified.

One approach that Griffith et al. might have used to clarify the meaning of *all* their constructs is to employ a table similar to Table 1.1 above. In the table, they could have shown the *things* that underlie each of their constructs (team or team member) and the attributes associated with the things. They also could have provided a rigorous definition of the construct. In the absence of Griffith et al. having defined all their constructs precisely, it is difficult to test their theory empirically. The reason is that valid and reliable measures cannot be devised for constructs that are not defined rigorously. Table 1 of their paper (p. 281) shows a number of constructs for which '[m]easures have to be developed', but valid and reliable measures cannot be developed unless the meaning of each construct is clear.

Associations

Griffith et al. state 19 propositions in their theory. Nine of these propositions manifest a single directional association between two constructs (five positive associations and four negative associations). Two propositions (P5a and P7) manifest two directional associations involving three constructs (one construct is associated with another construct that in turn is associated with another construct). Eight propositions manifest *moderated* associations—in other words, the strength of the directional association between two constructs is moderated by a third construct (an interaction effect is postulated).

Griffith et al.'s use of directional and moderated associations strengthens the potential predictive and explanatory power of their theory. Moreover, while they do not use the terms 'cause' and 'causal' when discussing their propositions, nonetheless causality is implied in the arguments they provide to support many of their propositions. For instance, it seems clear that they believe the existence of virtuality in a team *causes* certain outcomes to occur in relation to how different types of knowledge are transferred among team members. To the extent their propositions imply causality either implicitly or explicitly, the predictive and explanatory power of their theory is enhanced further.

Some arguments provided by Griffith et al. to support some of their associations are rigorous and compelling; however, two factors undermine the rigour of the arguments they use to support other associations. First, as discussed above, some of their constructs are not defined clearly. As a result, the meaning of any associations that employ these constructs will lack clarity. Second, because of the large number of constructs and associations employed in their theory, it is difficult to provide rigorous argumentation in support of all of them. Inevitably, some associations will be better argued than others.

Griffith et al.'s failure to specify all the associations in their theory rigorously undermines researchers' ability to test their theory empirically. In the absence of each association being articulated rigorously, researchers will lack the understanding they need to be able to evaluate whether the theory's associations hold empirically when they observe the outcome of a test of the theory.

Boundary

Griffith et al. do not use the term 'boundary' within their paper, nor do they have a specific section in their paper that discusses the boundary of the theory they are proposing. Nonetheless, at one point in their paper they indicate that their theory is not applicable to all kinds of virtual teams: 'This model is presented from the perspective of virtual teams where membership is relatively stable, but with members having interaction both within the focal team, as well as with collocated others' (p. 269).

Use of the evaluation framework motivates considerations of whether the theory is constrained in other ways. For instance: does it apply to all kinds of tasks that a virtual team with a relatively stable membership might undertake? Does it apply when the virtual team is made of members having substantial differences in culture? Does it hold throughout all phases of the virtual team's existence? Griffith et al. are silent on such questions. In the absence of their specifying the boundary to their theory clearly, however, researchers might test their theory in an inappropriate context.

Whole

In this subsection, I evaluate Griffith et al.'s (2003) theory as a whole. The evaluation of their theory's emergent attributes is more judgmental than the evaluation of their theory's parts.

Importance

In the introduction to their paper, Griffith et al. provide some clear and compelling reasons why the phenomena they are investigating are important for practice. They point out that the management of teams and knowledge is an important way of creating 'synergies in…resources' and 'increased value' for organisations (p. 266). Moreover, with the emergence and ongoing refinement and development of collaboration technologies and the increasing globalisation of workforces, virtual teams are becoming more prevalent. Thus, the successful operation of virtual teams is now critical to the success of many organisations (for example, Lowry et al., 2010).

From a research perspective, Griffith et al. argue their research potentially provides a foundation for other researchers who wish 'to identify the limiting conditions for effective learning and knowledge transfer across the range of traditional, hybrid, and virtual teams' (p. 280). It is clear that they have achieved this outcome, because Google Scholar shows their paper has been cited approximately 300 times.

Novelty

Prima facie, it does not appear that Griffith et al.'s paper has been paradigm changing in the sense that it has fundamentally altered the ways in which researchers view phenomena associated with virtual teams and knowledge transfer. In short, they are following a normal-science approach in their research (Kuhn, 1996). Nonetheless, their research can be deemed novel for several other reasons.

First, at the time their paper was published, their theory included a number of constructs that, if not completely new, had received only cursory attention in the extant research literature. For instance, Table 1 of their paper (p. 281) shows several constructs where they note 'measures to be developed'. Table 1.1 above also contains a number of constructs that, to the best of my knowledge, have not been canvassed extensively by researchers (for example, 'level of social interaction limited by virtual work undertaken').

Second, their paper included a number of associations that had received either no or only cursory attention in the research literature that existed at the time they prepared their paper. For instance, based on their analysis of the existing literature, the eight moderated associations they proposed in their theory appear to be new.

Third, the 'package' of constructs and associations included in their theory was novel. While at the time their paper was prepared other researchers might have canvassed *subsets* of the constructs and associations covered by Griffith et al.'s theory, the 'whole' was new. Their theory covered team virtualness and knowledge transfer phenomena in novel and potentially interesting and important ways.

In the context of Locke and Golden-Biddle's (1997) two strategies for demonstrating the contribution to knowledge of a piece of research, Griffith et al. first *construct inter-textual coherence* using Locke and Golden-Biddle's tactic of 'synthesised coherence'—making connections between literatures that historically have been somewhat disjointed (Locke and Golden-Biddle, 1997, pp. 1030–5). They enact Locke and Golden-Biddle's second strategy, *problematising the existing literature*, by using the tactic of 'incompleteness'—that is, showing that the existing literature can be characterised by knowledge gaps or lacunae (Locke and Golden-Biddle, 1997, pp. 1030–5).

Those tactics are manifested in the way they frame the contribution of their paper: 'The model is largely drawn from the extant literature...Our contribution is in combining the results from the prior literature in a way that is amenable to an assessment of the opportunities and challenges presented by considering more and less virtual teams from the perspective of knowledge' (p. 270). In short, Griffith et al. have tacitly followed Locke and Golden-Biddle's recommendations for demonstrating novelty and contribution via the rhetoric used to contextualise a piece of research.

Parsimony

As I indicated above, I believe Griffith et al.'s theory contains

- 29 constructs (rather than 17 constructs, as shown in Figure 2 of their paper, or 15 constructs, as shown in Table 1 of their paper)
- 19 associations (these are shown as propositions in their paper)
- One boundary condition.

Based on a simple count of the number of constructs and associations in Griffith et al.'s theory, it is difficult to conclude their theory is parsimonious. As a result, one might predict that this lack of parsimony would undermine the impact of the theory on other researchers. Interestingly, as indicated above, citation data suggest otherwise. Given the large number of citations of Griffith et al.'s paper, their theory clearly has had an impact on other researchers. Thus, contrary to expectations, lack of parsimony has not undermined the impact of their paper.

Level

In my view, Griffith et al. have articulated a middle-range theory. The range of phenomena that their theory covers is reasonably broad. Thus, they cannot be accused of narrow empiricism. Moreover, while a number of their constructs have yet to be defined rigorously and to be operationalised, it is possible to conceive how ultimately these outcomes might be achieved. In short, their theory is framed at a level that enables it to be employed to generate useful predictions and insights about, and understanding of, their focal phenomena.

Falsifiability

I have argued above that some parts of Griffith et al.'s theory have been articulated clearly and that other parts of the theory have *not* been articulated clearly. Where clarity exists, empirical tests can be undertaken to test the theory. Potentially, the outcomes of these tests will lead researchers to conclude that Griffith et al.'s theory is not supported. That is, the theory can be falsified. For those parts of their theory that are not articulated clearly, however, attempts to falsify the theory are problematic. Empirical tests that produce 'unfavourable' outcomes might simply mean that researchers have used invalid or unreliable measures for constructs or that they have failed to understand the nature of an association between constructs. They also might have tested the theory in a context that falls outside its boundary.

Conclusions

The framework I have proposed above facilitates an evaluation of the quality of an existing theory. It also informs researchers who are seeking to build a new theory or enhance or adapt an existing theory. As they construct their new or modified theory, they should be mindful of matters they need to address from the perspective of achieving high-quality outcomes in relation to the parts and whole of their theory. In essence, the framework can be used as a set of checkpoints to test the quality of the work they are undertaking.

The framework does not assist, however, in choosing the focal phenomena and the ways these phenomena might be conceived, nor does it assist in choosing a theory's constructs, associations and boundary. To a large extent, these choices remain creative acts that affect, in particular, the quality of the whole—a theory's importance, its novelty, its parsimony, and so on. In the information systems discipline (and in a number of other disciplines), I believe a rich vein of research lies in seeking to better understand the characteristics of those choices that have led to the articulation of high-quality, high-impact theories.

References

Boudon, R. (1991). Review: what middle-range theories are. *Contemporary Sociology, 20*(4), 519–22.

Bunge, M. (1977). *Treatise on Basic Philosophy. Volume 3. Ontology I: The furniture of the world.* Dordrecht, Netherlands: D. Reidel Publishing Company.

Bunge, M. (1979). *Treatise on Basic Philosophy. Volume 4. Ontology II: A world of systems.* Dordrecht, Netherlands: D. Reidel Publishing Company.

Doty, D. H., & Glick, W. H. (1994). Typologies as a unique form of theory building: towards improved understanding and modeling. *Academy of Management Review, 19*(2), 230–51.

Dubin, R. (1978). *Theory Building*. [Rev. edn]. New York, NY: Free Press.

Eisenhardt, K. (1989). Building theories from case study research. *Academy of Management Journal, 14*(4), 532–50.

Gregor, S. (2006). The nature of theory in information systems. *MIS Quarterly, 30*(3), 611–42.

Griffith, T. L., Sawyer, J. E., & Neale, M. A. (2003). Virtualness and knowledge in teams: managing the love triangle of organizations, individuals, and information technology. *MIS Quarterly, 27*(2), 265–87.

Kuhn, T. S. (1996). *The Structure of Scientific Revolutions*. [Third edn]. Chicago, Ill.: University of Chicago Press.

Lewin, K. (1945). The research center for group dynamics at Massachusetts Institute of Technology. *Sociometry, 8*(2), 126–35.

Locke, K., & Golden-Biddle, K. (1997). Constructing opportunities for contribution: structuring intertextual coherence and 'problematizing' in organizational studies. *Academy of Management Journal, 40*(5), 1023–62.

Lowry, P. B., Zhang, D., Zhou, L., & Fu, X. (2010). Effects of culture, social presence, and group composition on trust in technology-supported groups. *Information Systems Journal, 20*(3), 297–315.

Markus, & Robey, D. (1988). Information technology and organizational change: causal structure in theory and research. *Management Science, 34*(5), 583–98.

Merton, R. K. (1957). *Social Theory and Social Structure*. Glencoe, Ill.: Free Press.

Miller, G. A. (1956). The magical number seven, plus or minus two: some limitations on our capacity for processing information. *Psychological Review, 63*(2), 343–55.

Mone, M. A., & McKinley, W. (1993). The uniqueness value and its consequences for organization studies. *Journal of Management Enquiry, 2*(3), 284–96.

Popper, K. (1990). *The Logic of Scientific Discovery*. London, UK: Unwin Hyman.

Weber, R. (2003). Theoretically speaking. *MIS Quarterly, 27*(3), iii–xi.

Zmud, R. (1998). Editor's comments. *MIS Quarterly, 22*(2), xxix–xxxi.

Zmud, R. (2003). Special issue on redefining the organizational roles of information technology in the information age. *MIS Quarterly, 27*(2), 195.

2. Obstacles to Building Effective Theory about Attitudes and Behaviours Towards Technology

Mary Tate
Victoria University of Wellington

Joerg Evermann
Memorial University of Newfoundland

Abstract

Some widespread approaches and practices in information systems research might be doing more harm than good, and potentially acting as obstacles to effective theory building. Theories of attitudes and behaviours towards technology have formed part of the core of the information systems discipline for 30 years, but, despite this, there is still relatively little consensus about the definition, operationalisation and nomological net of many key constructs. There are also ongoing calls for theory that is more salient to practice. In this conceptual chapter, we identify 10 potential obstacles to effective theory building that could occur during presentation of previous literature, theory development and research design. We define each potential obstacle, explain why it is a problem and offer some alternatives.

Introduction

Ever since the days of Keen and Weber, influential thinkers within the information systems (IS) discipline have bemoaned the lack of a cumulative research tradition (Keen, 1980; Weber, 1997). For the discipline to progress, it has been suggested, IS research needs fewer frameworks and more ongoing cumulative lines of inquiry (Keen, 1980; Weber, 1997).

More recently, some have argued that the academic legitimacy of the field does not necessarily depend on having a core of theory, but on the strength of results and their salience to practice (Lyytinen and King, 2004). Despite this, there continues to be a strong emphasis, especially in our top journals, on the importance of the theoretical contribution of research: 'The contribution of the paper is its theory (backed by the evidence), not the empirical evidence itself (however interesting that may be)' (Seddon, 2006, p. 5).

A core component of the body of IS theory is theories of attitudes and behaviours towards technology, such as the technology acceptance model (TAM) (Davis et al., 1989), the task-technology fit model (Goodhue and Thompson, 1995) and the service quality of the IS function (Pitt et al., 1995). Despite 30 years of research, some theories of attitudes and perceptions towards technology are characterised by inconsistent, even conflicting, definitions and operationalisations, while others—considered more stable—have made a limited contribution to knowledge. As an example of the first, the definition of online service quality in different contexts has included 'overall affect towards a web portal' (Yang et al., 2005), 'intent to repurchase' (Sum et al., 2002) and 'interactions between citizens and government officials' (Teo et al., 2009). As an example of the second, the core constructs of the TAM (perceived usefulness and perceived ease of use) have been reasonably consistent but the overall theoretical contribution of the accumulated body of TAM research has been criticised on a number of counts (Benbasat and Barki, 2007). These include: a bewildering number of extensions with marginal incremental value; little overall insight for our practitioner community into the design factors that lead to usefulness and ease of use; and the possibility that the dominance of TAM has constrained the inclusion of other relevant beliefs in our theories (Benbasat and Barki, 2007).

Given this situation, it is possible that some aspects of the way we commonly conduct theory development and research design would benefit from critical evaluation. In this chapter, we present a set of potential obstacles to effective theory building in this area of information systems. We have concentrated on quantitative methods, as these are the most widely used in studies of attitudes and behaviours towards technology in top-ranked journals. While each issue has an extensive literature in its own right, our intention is to provide a broad view of the potential impact of these issues on IS theory building, rather than an in-depth discussion of each issue. A further aim of this chapter is to introduce literature from reference disciplines, which offers a more in-depth discussion of these issues.

Some of our assertions run counter to what might be termed received wisdom in our field, and contrary views are possible and indeed welcome. Overall, we wish to stimulate debate about these important issues in the hope that this discourse will contribute to an overall improvement in the quality of theory building in the discipline. We have divided the issues into three sections: those associated with presenting previous literature, those associated with theory building and those associated with research design. We briefly define each issue and explain why it is a problem for effective theory building, and offer some examples and alternatives.

Obstacles in Presenting Previous Knowledge

Literature reviews have 'played a decisive role in scholarship' because 'science is a cumulative endeavour' (vom Brocke et al., 2009, p. 1). The literature review is the artefact that enables the researcher to identify the gap in knowledge that their research will address and to distinguish the original contribution of the research from what has gone before. The literature review underpins the development of new theory. This means that issues with the presentation of previous literature can act as constraints on theory building.

Before identifying potential problems, we first examine different types of literature reviews. Figure 2.1 shows four methods of literature review: narrative review, descriptive review, vote counting and meta-analysis. These four review methods are placed in a qualitative–quantitative continuum to illustrate their different foci (King and He, 2005). The 'narrative review' is the traditional way of reviewing research literature. It is conducted by verbally describing past studies, and focusing on theories and frameworks, elementary factors and their research outcomes, with regard to arriving at hypothesised relationships (King and He, 2005) (Figure 2.1).

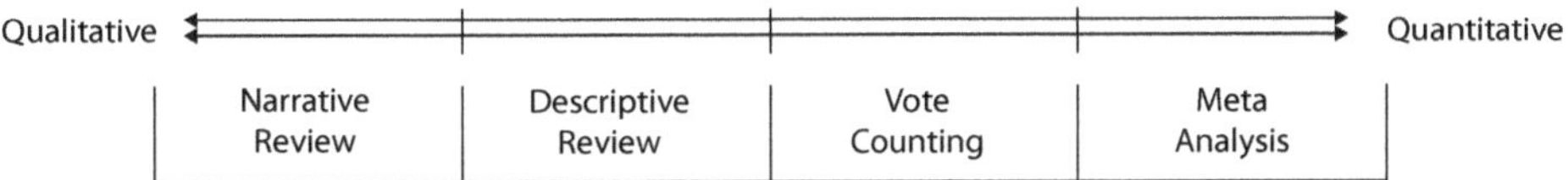

Figure 2.1 Literature Review Methods on a Qualitative–Quantitative Continuum

A descriptive review focuses on revealing an interpretable pattern from the existing literature (Guzzo et al., 1987). It produces some quantification—often in the form of frequency analysis, such as publication time, research methodology, research outcomes, and so on. Such a review method often has a systematic procedure including searching, filtering and classifying processes. The reviewer treats an individual study as one data record and identifies trends and patterns among the papers surveyed (King and He, 2005). The outcome of such a review is often claimed to be representative of the state of the art of a research domain.

Vote counting is generally used to draw inferences upon focal relationships by combining individual research findings (King and He, 2005). Here a tally is made of the frequency with which existing research findings support a particular proposition. Most likely it is applied to generate insights from a series of experiments. The premise underlying this approach is that repeated results in the same direction across multiple studies—even if some of them are non-significant—might be more powerful evidence than a single significant result (King and He, 2005). We were not able to identify any examples of this type of literature review in the IS area.

Meta-analysis aims at statistically providing supports to a research topic by synthesising and analysing the quantitative results of many empirical studies (King and He, 2005). In most cases, it might specifically examine the relationships between certain independent variables (IVs) and dependent variables (DVs) derived from existing research findings. Qualitative studies have to be excluded from a meta-analysis due to its extremely quantitative nature. Only studies with comparable effect size metrics can be included in a meta-analysis.

Of these types, the narrative literature review, which aims to synthesise previous knowledge, is the most commonly used form in IS research papers. In this section, we examine some obstacles that can arise from the approach and stance adopted by the researcher to presenting previous research.

The Unconsciously Subjective Literature Review

Description

Literature reviews are almost always a form of qualitative research. A narrative literature review in particular is a qualitative and subjective review that, far from being value neutral, requires and invites the researcher to adopt a distinct viewpoint. The subjective influences stem from a consideration of what literature sources (journals, conferences, range of publication dates) are of sufficient importance to consider, which theories provide sufficient evidence to be trusted and which methods are considered sufficiently rigorous to yield valid results (vom Brocke et al., 2009). There are few reviews that make explicit all assumptions underlying the scope and selection of studies or that employ multiple researchers to alleviate individual biases. This means that the researcher adopts a value-neutral stance while making unconscious (or unacknowledged) subjective judgments.

Narrative literature reviews are subjective in the identification of concepts and relationships, the establishment of their equivalences across studies and the assessment of validity. All researchers have subjective positions on these and related questions, of which they might not even be explicitly aware. The narrative literature review thus necessarily represents a qualitative and subjective interpretation of the literature. It is therefore not unusual for 'two reviews to arrive at rather different conclusions from the same general body of literature' (Guzzo et al., 1987, p. 408).

Rigorous criteria for selection and evaluation of literature are mostly reserved for review articles, while the literature review for 'regular' articles does not usually exhibit these characteristics. Descriptive literature reviews might be only slightly less subjective, but the subjective criteria are more explicit. Descriptive literature reviews require the researcher to articulate a systematic procedure for searching, filtering and classifying research articles.

Why is This A Problem, and What Should We Do Instead?

In the IS field, a number of constructs of interest—such as service quality, diffusion of innovation and technology adoption—have been characterised by a complex and heterogeneous literature. It is possible to find reputable, previously published studies backed by empirical studies that show that A causes B, and equally reputable studies, also backed by empirical evidence, that find just the opposite. A literature review that uncritically accepts both positions without being able to explain the difference in findings is worse than useless. It signals to the reader and future researchers that there is no problem in this area of research—that somehow the theories might be commensurable, usually by ensuring that the concepts are sufficiently wide or elastic that contradictory results are possible.

Moving to a descriptive review or a meta-analytic study will assist this issue. A descriptive review needs to articulate the criteria adopted for the selection and interpretation of the studies. Hence, being sufficiently explicit about the relevant literature sources, and being explicit not only in what to include but also in what to exclude, and why, will allow researchers to come to a consensus on the state of the art in some research area.

For a meta-analytic review, these criteria must include not only the desirable properties of theories (the ability to explain and predict, parsimony and coherence with other theories), but must also extend to the method employed by prior studies. Rarely are the results and research methods of prior work critically examined based on *current* standards of validity. There is no point including flawed studies in a meta-analysis. Frequently, the details of previous studies are assumed to be correct. After all, the paper has been published, perhaps in a 'good' journal; however, criteria for validity are not fixed over time and what was acceptable 20 years ago might not be acceptable anymore. For example, exploratory factor analysis (EFA) is not as acceptable as it was 20 years ago, before the widespread availability of structural equation modelling, and many now warn against reliance on exploratory techniques for various reasons. Hence, it behoves researchers to reassess EFA-based results, rather than reusing them indiscriminately. An extended debate on this topic can be found in a special issue of *Multivariate Behavioral Research* (volume 31, issue 4) (David, 1996; McDonald 1996; Maraun, 1996a, 1996b, 1996c; Mulaik, 1996; Rozeboom, 1996; Schonemann, 1996a, 1996b; Steiger, 1996a, 1996b).

In summary, since the literature review is inherently subjective and cannot be value neutral, it is important that this process is as transparent as possible, and is based on explicitly acknowledged criteria for the selection and evaluation of articles. This includes a critical examination of the method applied according to current standards and heuristics in research methodology.

Meaning Variance: Does a rose by any other name smell as sweet?

Description

Sometimes, the same term is used for constructs that are measured with different, non-equivalent items, while different terms are used for constructs that include some of the same items (Barki, 2008). These ambiguous constructs are then included in different theory nets. This tends to create and compound problems with meaning variance. Meaning variance occurs when the same term is used but it is defined or operationalised differently, so the constructs being synthesised or compared are not equivalent. This has problems for theory building as researchers can 'utter the same words, but the words have different meanings, so any logical comparison of their utterances is precluded, leaving adherents of rival theories simply talking past one another' (Curd and Cover, 1998, p. 222).

By way of example, a paper by Petter et al. (2008) provides a narrative literature review where the equivalence of different constructs is argued by the authors, and could be open to other interpretations. They argue that constructs identified as instances of 'service quality' incorporate other quite disparate constructs, including '[t]he effective role of the technical staff (service quality)'; 'the retention of service staff (and the related funding)' and the 'competency of support staff, vendor support, and availability of training' (Petter et al, 2008, p. 245).

Many arguments and counterarguments have been proposed concerning the definition and operationalisation of service quality, particularly the ServQual instrument, variations of which are used for measuring customer perceptions of service quality in a range of contexts. According to the Google Scholar search engine, ServQual (Parasuraman et al., 1988) has been cited more than 6000 times.

Despite the popularity of ServQual, many reputable studies have failed to replicate the five dimensions of service quality posited by the original authors (Babakus and Boller, 1992; Carman, 1990; Parasuraman et al., 1991). Although this issue is not fully resolved, it is frequently glossed over. ServQual is frequently perceived as 'the cornerstone on which all other works [about service quality] have been built' (Sureshchandar et al., 2002, p. 10).

Why Is This A Problem, and What Should We Do Instead?

The narrative literature review, as a qualitative, interpretative and subjective activity, allows the researcher much freedom in interpreting the input texts as well as in the production of the output text (the review). Thus, when the review does not provide a detailed examination of different concepts and their relationships, but instead glosses over any differences or particularities of their definition and operationalisations, the result is likely a conceptual mash-up that is overly broad in scope and includes related, but clearly distinguishable concepts. Moreover, the boundaries of the concept might become unclear or 'fuzzy'. When such fuzzy concepts are used within a nomological network of a theory, they might lead to tautologies, the inability to generate precise hypotheses or the inability to generate hypotheses that are unique to that theory. Moreover, these fuzzy concepts are then promoted to future researchers as state of the art, thus perpetuating the mistakes of the original review and endangering subsequent research.

As an alternative, a more nuanced treatment of discourse, which aims to identify the important areas of difference, rather than the points of similarity, is provided by Sylvester and Tate (Sylvester and Tate, 2008; Sylvester et al., 2007). These authors offer a representation of the richness of the service quality discourse, using soft-systems analysis and rich pictures (Sylvester et al., 2007). In addition, a critical discussion of the social and environmental factors associated with the discourse provides further insights, such as the importance of the determined ongoing advocacy of the original ServQual authors in developing and perpetuating the ServQual stream of research (Sylvester et al., 2007).

Overall, there are a number of research areas within information systems that are contradictory and heterogeneous, and thus not truly cumulative. Attempting to synthesise them glosses over important areas of difference and this tends to have the effect that the same term is used relatively indiscriminately to encompass an increasingly broad and vague set of definitions and operationalisations. A critical analysis or a more nuanced discussion that focuses on the diversity of opinions (Cooper, 1988; Killduff and Mehra, 1997; Rosenau, 1992) might be more appropriate in these circumstances.

Using Literature as Lego Blocks

Description

Coverage of previous literature is usually assumed to mean some kind of synthesis. Novice researchers are exhorted to integrate (synthesise) previous literature rather than simply summarise one study after another. Thus, recommendations of concept-centric rather than study-centric reviews abound (Webster and Watson, 2002). Furthermore, literature reviews exhort the researcher to find gaps in the theories they examine in order to motivate future research. This is further exacerbated by the insistence of reviewers and editors that a theory to be tested (and published) must be based on the literature. To the inexperienced and novice researcher, this means that 'combining the boxes' appears a safer method than introducing new and original thought.

Why Is This A Problem, and What Should We Do Instead?

In the most naive case, this procedure leads to simply 'combining the boxes'. Theories are developed by their combination or by addition of concepts, typically at the periphery of the theory, rather than by changes or redevelopment.

This process of playing Lego blocks with concepts and theories leads to overly simplistic extensions and work that is conservative rather than thought-leading or challenging. Hence, the results are theoretically sound, but applicable only to the specific area for which a theory extension was proposed. This is how the IS field ends up with dozens of extensions of TAM into various niche fields, as well as hundreds of different operationalisations of TAM for different phenomena. It is interesting to note that the majority of these are not actually published in the top-tier journals, but, due to their conservative nature, are relegated to second-tier outlets.

Instead of playing Lego blocks with theories and concepts, researchers should identify gaps in knowledge by what is left unexplained by existing theories. These 'mysteries' might not only require extension of theory, they might also lead to re-conceptualisation of an entire area and thus produce interesting and challenging new theories. These paradigm shifts have occurred in many other areas of science but are relatively rare in information systems. Examples might be to examine organisations not as systems of control and input–output, but as 'orderly arrangements of individual human interactions' (Tannenbaum, 1962, p. 236) or other metaphors (Evermann, 2005a; Reed, 1992; Walsham, 1991). System implementation projects can also be conceptualised using metaphors such as organisational drift and organisational politics, rather than the more common conceptualisation as a set of activities to be optimised (Hirschheim and Newman, 1991; Kendall and Kendall, 1993; Kling and Iacono, 1984).

Reference Disciplines Lost in Translation

Description

Information systems has a wide range of reference disciplines, and many areas of inquiry in IS can be considered as special cases of wider areas of inquiry (Baskerville and Myers, 2002). For example, the formation of attitudes and beliefs towards technology can be considered as a special case of the formation of attitudes and beliefs in general—the domain of social psychology. The study of IS service quality was based extensively on the study of service quality by marketing researchers.

The issues with our reporting of knowledge from reference disciplines are threefold. First, there is a danger of being 'lost in translation', where details and nuances of original theory are lost in the process of appropriation by IS authors. The second is the time lag that frequently exists from the reference discipline to publication within the IS literature. Frequently, IS researchers do not acknowledge or cite new knowledge from reference disciplines until it has been published in our own leading journals. Then, once it has been published in our own leading journals, we often stop at that point in development, and do not reference later developments within the reference discipline. Finally, we do not critically re-examine previous IS studies in the light of new insights from reference disciplines. We illustrate this issue with the example of psychometric literature, the use of Bunge's ontology in design science research and studies in cognitive dissonance.

Why Is This A Problem, and What Should We Do Instead?

Attitudes and perceptions—including therefore attitudes and perceptions towards technology—are psychological states. Psychometrics is a reference discipline used by IS researchers, which concerns itself with the operationalisation and measurement of psychological states. A widely used approach for modelling and measuring psychological states is latent variable theory (Borsboom, 2005; Borsboom et al., 2003).

Latent variable theory—the understanding of formative and reflective latent variables, how they should be modelled and what validity criteria to apply—has only been discussed recently in IS literature (Petter et al., 2007), despite being widely discussed in psychometric literature for more than 10 years (see, for example: Diamontopoulos, 2006; Diamontopoulos and Winklhofer, 2001; Edwards and Bagozzi, 2000; Jarvis et al., 2003).

This is a problem because the modelling and interpretation of formative latent indicators, and evaluations of model quality and fit, are quite different to that for reflective latent variables (Chin et al., 1988; Diamontopoulos and Winklhofer, 2001; Petter et al., 2007). Researchers who are unaware of this distinction run the risk of seriously mis-specifying their models. Jarvis et al. (2003) have shown that being unaware of the distinction can have dire consequences for the resulting estimates. Furthermore, widely cited models that were published before this knowledge became mainstream in IS might be mis-specified (Tate and Evermann, 2009a).

Overall, theories of quantitative psychological measurement methods do not form part of the core of the IS discipline, but theories of attitudes and behaviour towards technology do; however, these theories depend heavily on methods, metrics and heuristics developed in psychometrics. It is likely that we will continue to need to look to psychometrics for guidance as to how best to conduct psychological measurement.

A second example might be the adoption of Bunge's ontology to ground much of current design science research. It should be noted that Bunge's ontology is an eight-volume work, yet Wand and Weber (1993) have limited their adoption to about half of the first volume, claiming that this is all that is required for IS researchers, and adapting and changing the ontology in the process. Moreover, other ontologies exist that might be as appropriate or more so (Milton and Kazmierczak, 2004). While some argue that the post-hoc empirical success of this ontology in explaining different phenomena around IS design justifies the adoption of and reliance on this particular adaptation of this particular subset of this particular ontology, these studies often have challenges in validity (Evermann, 2005b). For example, rarely is the underlying theory sufficiently formalised and sufficiently specific to exclude competing explanations of the phenomena. Different authors have at times challenged the use of this ontology as inadequate (Evermann, 2005b; Wyssusek, 2005), and other areas of science have developed a very different understanding of ontology than the IS field (Noy and Hafner, 1997; Uschold and Gruninger, 1996).

Despite the increasing maturity of IS as a discipline, it behoves IS researchers to be cognisant of ongoing developments in reference disciplines, rather than relying exclusively on the most recent 'IS translation' of the theory published in a leading IS journal.

A final example is cognitive dissonance theory. Based on original work by Festinger (1957), it has been considerably refined in a process of cumulative studies since then (for a summary of this development, see Cooper, 2007). Yet, even recent IS studies that are based on cognitive dissonance make reference only to Festinger's original work (Bhattacherjee and Premkumar, 2004; Pavlou and Gefen, 2005). Important limitations or re-conceptualisation in light of self-image and self-perception theory are omitted or glossed over. For example, Festinger's (1957) theory of cognitive dissonance survived many challenges but has been transformed and re-conceptualised beyond what Festinger imagined in 1957. Such transformations might make the theory inapplicable in some of the situations that Bhattacherjee and Premkumar (2004) and Pavlou and Gefen (2005) describe.

When making use of reference disciplines, a literature review must include more than the original study. Theories are not static constructs and reference disciplines extend, constrain, re-conceptualise or otherwise adapt their theories over time. Thus, while researchers might begin with classic or seminal papers, a forward literature review is absolutely necessary. Tools such as the ISI Web of Science make forward literature reviews much easier than they used to be and leave researchers little excuse for not performing them.

Obstacles in Theory Development

There is a tension in theory development between accuracy, simplicity (parsimony) and generalisability, which are all highly valued qualities of theory. One effect of this is a tendency to focus on the more generalisable, but less accurate and salient aspects of attitudes and behaviours towards technology. Information systems researchers have sometimes been wary of engaging deeply with specific technologies as these are seen as transient phenomena; however, it is precisely in understanding and explaining the interface between people and technology artefacts that IS research has the potential to make a unique contribution (Benbasat and Zmud, 2003). In this section, we examine the issues surrounding the way we balance these three goals, and offer some suggestions as to how this could be improved.

Valuing Parsimony Over Accuracy

Description

Parsimony is frequently considered a characteristic of quality in a theory. There is, however, an inherent tension between theory parsimony and a nuanced and accurate description of the phenomenon of interest.

Parsimony is only one desirable aspect of theory. Accuracy, consistency and scope (generalisability) are also highly valued (Kuhn, 1983). Another important point that should be added is originality, which is a key publication criterion for top IS journals (Seddon, 2006). There are inherent tensions between accuracy, generalisability and simplicity (parsimony) (Figure 2.2).

This tension is not unique to IS, but is an ongoing issue within social science research (Blalock, 1982). The diversity of social phenomena tends to lead to issues with theory generalisability and parsimony (simplicity). The quest for parsimony is often the cause of a gap between models and the complexities of the real world. Although social scientists

> strive for theories that are simultaneously parsimonious, highly general, and therefore applicable to a wide range of phenomena, yet precise

> enough to imply rejectable hypotheses, it does not appear possible...to achieve simultaneously all three of these ideal characteristics...my own position is that of the three, parsimony is the most expendable. (Blalock, 1982, p. 28)

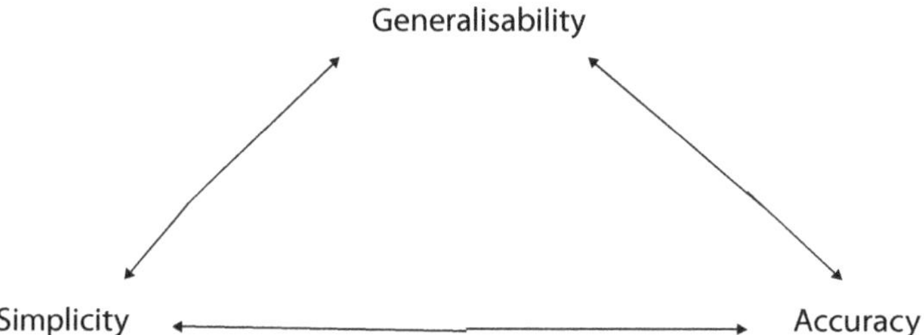

Figure 2.2 The Trade-Off Between Generalisability, Accuracy and Simplicity in Theory

Why Is This A Problem, and What Should We Do Instead?

In 1998, Robey and Marcus lamented that IS theory is not 'well positioned to recommend actions for improving the intervention' of a new technology (Robey and Markus, 1998, p. 10). Ten years later, theories of user attitudes and perceptions towards technology are still 'leading to research that is unable to provide actionable advice' (Benbasat and Barki, 2007, p. 213).

We contend that the pursuit of parsimony is at least partly to blame. To illustrate this, we compare two widely cited papers on IS success. According to Google Scholar, DeLone and McLean's (1992) IS success model has been cited more than 3000 times, including being extended and applied in multiple business contexts. It is a relatively simple model, with seven constructs, each of which, the authors suggest, could be measured (operationalised) by a set of three to five metrics. The elegance and simplicity of this model are almost certainly a major part of its appeal.

By way of contrast, we consider the critiques and alternative proposed by Ballantine et al. (1996) in their '3-D' model. Among the critiques they make of the original DeLone and McLean model are the fact that it focuses too heavily on financial aspects and does not consider human aspects, that it fails to recognise that different factors contributing to success might be carried out at different levels of the organisation and the organisational structure of the organisation might impact on IS success, that it does not take into account the iterative nature of IS systems and the degree to which learning might occur during their life cycle, the impact of the external business environment, the effectiveness of the organisation's change processes and other information systems. The resulting 3-D model attempts to incorporate these factors.

The 3-D model is complex, with perhaps 40–50 constructs, organised in several layers, with various 'filters' and external ('exogenous') factors acting on them. It is a much more complete and accurate picture of the complexities of IS management and measurement in an organisational context than the DeLone and McLean model; however, the 3-D model would be almost impossible to operationalise and test quantitatively. It is anything but parsimonious. It is also far more likely to be useful and salient. For example, Ballantine et al.'s model would be far more useful as a guide for a post-implementation review of a recently implemented system than the more parsimonious IS success model.

That the quest for parsimony has led IS researchers to make dubious claims of explanatory success has been shown by Evermann and Tate (2009a) who, upon examining recent studies using structural equation analysis, conclude that most of the published theories do not fit the data, despite claims to the contrary by the authors. They have shown in one particular instance that the search for parsimony is to blame and that a more nuanced theory is in fact able to explain the observed data.

Overall, there is a tension between parsimonious models, which are inherently reductionist, and information systems, which as social phenomena are inherently complex. Salience for practice is likely to require a richer and more holistic understanding of the phenomena than our parsimonious models typically provide.

Excessive Claims of Generalisability

Description

There has been ongoing dialogue within the IS community about the degree to which IS researchers should aim for generalisability of theory and the tension between accuracy, simplicity and generalisability (Seddon and Lyytinen, 2008). As we discussed, there is a trade-off between generalisability, accuracy and simplicity (Figure 2.2). Within that, generalisability as well as simplicity (discussed above) is often privileged at the expense of accuracy.

One reason generalisability is valued in theory development is that it is assumed to be evidence of the development of a cumulative research tradition. Generalisable models might be created by expanding the scope (boundaries) of existing models or by combining two or more models with a more limited scope (Dubin, 1978).

Why Is This A Problem, and What Should We Do Instead?

In their recent discussion (Seddon and Lyytinen, 2008), the panellists noted that many studies are not as representative as they claim to be, because the technologies and users that are studied vary considerably. For example, not all email or enterprise resource planning (ERP) systems are the same, and the characteristics and perceptions of individual users can vary within and between different contexts of use.

In the IS research community, issues have been raised about the state of research into user attitudes and perceptions towards technology in a special issue of the *Journal of the Association of Information Systems* (Benbasat and Barki, 2007; Hirschheim, 2007). Benbasat and Barki (2007) suggested that the popularity of TAM has diverted researchers away from its antecedents—in particular, the design of the IT artefact and the characteristics that make it useful. They further suggest that the addition of various constructs (other models with a different scope) such as trust and self-efficacy to TAM has created the 'illusion of cumulative tradition', but has not in fact extended the boundaries of the theory (Benbasat and Barki, 2007, p. 213).

As an alternative, we can both test the generalisability of some of our core constructs and improve the salience of our models for practitioners, by extending our models of attitudes and perceptions towards technology to include more consideration of the underlying technological artefact. For example, there is currently a great deal of competition in the social networking space. All leading social networking sites with large established user bases presumably meet basic expectations of usefulness, ease of use and task-technology fit, or adoption would not have been so extensive. Competition is enacted at a much more detailed level, such as (for example) the visibility, attractiveness and placement of social network software bookmarking icons. Evermann and Tate (2009b) provide an example of how the perceived presence of a social networking site bookmark can be modelled using antecedents such as visibility, which in turn has very specific antecedents such as placement and size. While it is true that this is constrained to a relatively narrow phenomenon—social networking bookmarking icons—the value of social networking applications is currently in the billions of dollars per year, so research that leads to small gains in competitiveness is highly salient.

Overall, very small differences in technology affordances, functions and interactions might have a significant impact on competitiveness. Information systems theory should enable us to distinguish and measure those differences that are important, rather than concentrating exclusively on those aspects that are generalisable.

Failing to Distinguish Between Beliefs, Attitudes, Intentions and Behaviours

Description

Information systems theory draws heavily on theories of attitudes and behaviour from social psychology. For example, TAM (Davis et al., 1989) and its successor, UTAUT (Venkatesh et al., 2003), both cite the theory of reasoned action (TRA) (Fishbein and Ajzen, 1975) and its successor, the theory of planned behaviour (TPB) (Ajzen, 1991), as important parts of their provenance. Although the core theories are widely known, the detailed discussion that lies behind these theories is not.

Fishbein and Ajzen (1975) distinguish between affect, cognition and conation. Affect is a person's feeling towards and evaluation of some object (person, issue or event); cognition refers to his or her knowledge, opinions, beliefs and thoughts about the object; and conation refers to his or her behavioural intentions. Fishbein and Ajzen use 'attitude' to denote affect, 'belief' to denote cognition and 'intention' to denote conation (Figure 2.3). Beliefs link an object to some attribute, which might be any trait, property, characteristic, affordance or outcome associated with that object.

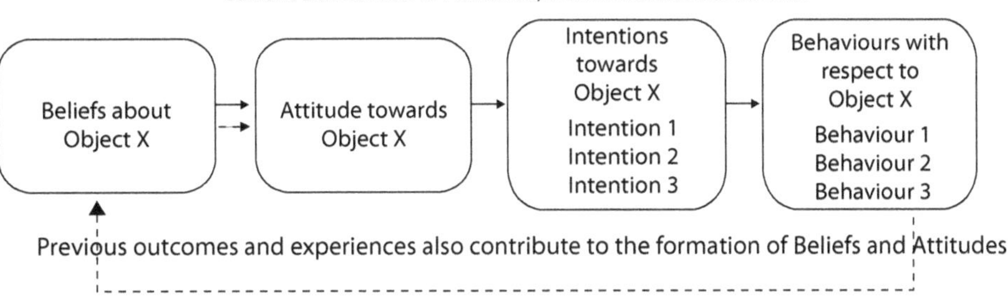

Figure 2.3 Beliefs, Attitudes, Intentions and Behaviours

Fishbein and Ajzen assert that the belief concept is central to understanding attitudes, intentions and behaviours. Beliefs about an object form the basis for the formation of attitudes towards that object. Belief formation involves the establishment of a link between two aspects of an individual's world. One is direct observation via the senses of an (technological) object (for example, that a web site offers a certain information item or button to push). This gives rise to descriptive beliefs about the object. Since people rarely doubt the validity of their own senses, descriptive beliefs are usually held with maximum certainty. Beliefs that go beyond directly observable events are inferential beliefs. These are formed from descriptive beliefs—for example, the presence of certain information might be used to infer usefulness.

Inferential beliefs might be based in the first instance on previous inferences, but Fishbein and Ajzen claim that most inferential beliefs can eventually be traced back to descriptive beliefs. The distinction between descriptive and inferential beliefs is, however, a continuum rather than a dichotomy:

> At the descriptive end of the continuum, a person's beliefs are directly tied to the stimulus situation, and at the inferential end, beliefs are formed on the basis of these stimuli as well as the residues of the person's past experiences; the continuum may be seen as involving [a range from] maximal to minimal use of such past experiential residues. (Fishbein and Ajzen, 1975, p. 133)

Why Is This A Problem, and What Should We Do Instead?

Information systems theories of perceptions and attitudes towards technology have often failed to adequately distinguish between descriptive and inferential beliefs. The terms 'perception', 'user's perception' or 'customer's perception' are often used in the IS and consumer behaviour literature as catch-all terms that encompass both descriptive and inferential beliefs. For example, the perceptions of web site users of the navigability of a site are likely to be more descriptive than inferential: they are reasonably precise and likely to be well correlated to objective features such as the number of clicks required to achieve a goal. There might be some variation based on the individual, but the responses are likely to be reasonably similar between individuals, with a substantial portion of the differences in score explained by objective qualities of the site.

The belief of whether a site is easy to use, however, is a more inferential than descriptive belief. There are two issues with inferential beliefs that place them at a remove from salient features of the technology. The first is that they are much more generalised than descriptive beliefs, which greatly reduces their salience in any specific context (Tate and Evermann, 2009b). The question 'is it easy to use' is a question one could reasonably ask about a diversity of technologies, ranging from lawnmowers to surgical instruments; however, if we do not include the antecedent descriptive beliefs, which are based on the affordances of the technology then this question, although it might have the same wording, is effectively a different question when used in different contexts. The characteristics and features that contribute to respondents rating a lawnmower as 'easy to use' might include how easy it is to start, its weight and manoeuvrability, and so on, and will be different to those that contribute to the evaluation of electronic surgical instruments (for example, accuracy, visual quality and response latency). Measuring the inferential beliefs alone does not provide any clues to inform the design of either technology.

The other issue with inferential beliefs and the subsequent formation of attitudes towards a technology is that they tend to accumulate over time and with experience (Tate and Evermann, 2009b). For example, consider two users who are asked to provide their perceptions of the trustworthiness of an e-commerce web site, and then to indicate their attitudes and intentions towards conducting e-commerce transactions on that site. One user has previously experienced credit card fraud on the Internet (but not on that site) and one has not. They will likely have similar descriptive beliefs as to whether the site offers specific features shown to be antecedent to trust in electronic commerce such as privacy and security (Pittayachawan et al., 2008); however, the user who has previously experienced credit card fraud might hold a lower degree of trust (an inferential belief) and a more negative attitude to transactions on that web site, and to e-commerce in general, because they have an experiential residue from their previous fraud experience. These unmeasured causes make the measurement of inferential beliefs and attitudes less reliable.

Overall, by understanding and explicitly modelling the differences between beliefs, attitudes, intentions and behaviours, we can create more accurate models about the formation of attitudes and intentions towards technology. In particular, we can identify beliefs about specific technology affordances as antecedent to the formation of attitudes, which has the potential to result in more accurate and salient theories.

Obstacles in Research Design

Our final set of obstacles relates to our choice of research designs to operationalise and evaluate our theories. Human attitudes and behaviours towards technologies are inherently complex social phenomena that lend themselves to multiple interpretations and explanations. We like to believe that there could be simpler sets of underlying factors that give rise to and explain some of this complexity. Hence, data reduction techniques such as factor analysis are popular, but potentially misleading. We also like to think that a carefully designed study, with a large sample and that produces results with good fit to our model, has discovered something meaningful about the world. This might be the case, but we suggest that we are in danger of moving too quickly and easily from one well-designed and executed study to an assumption of 'fact'. We seldom replicate studies, examine alternative models or examine the possibilities arising from unexpectedly positive or negative results.

Inappropriate Use of Factor Analysis

Description

Factor analysis is frequently used inappropriately to 'discover' psychological constructs. This description is summarised from a longer and somewhat more technical discussion in a special issue of *Multivariate Behavioural Research*, which was devoted to discussion of factor analysis.

Maraun (1996b) identifies two fallacies that are often associated with the use of exploratory factor analysis (EFA). The first is the assumption that the common factors of factor analysis are necessarily underlying, hidden, unmeasurable (directly) or unobservable (directly) variables. The second is the assumption that factor analysis detects the existence of these variables and their influence on the measurable variables (indicators, survey questions, survey items).

In reality, common factors identified in EFA are none of these things. This is because in EFA the proof of the 'existence' of the factor is internal to the analysis, and factors are constructed by the analysis, not hypothesised *a priori*. According to Maraun (1996b), claims that factor analysis deals with some other factor variate—for example, a hypothetical (existing external to the data) variate—are 'confused'.

Why Is This A Problem, and What Should We Do Instead?

When EFA claims to have 'discovered' psychological latent factors, such as attitudes towards technology, the exploratory factor analysis is doing double duty. The same set of calculations that constructs the factors in the first place cannot also be used as evidence for their independent existence in the minds of the respondents. The researcher cannot, in the same breath, build or construct something and claim to have discovered it. This is a fallacy known as factor reification.

The impact of this is that many models contain constructs 'discovered' by factor analysis to which researchers then apply names and definitions, and specify reflectively (antecedent to their indicators). Because EFA merely decomposes the matrix of *undirected* correlations it cannot distinguish between antecedents and consequents. Hence, many of these constructs are not in fact reflective latent variables, but composites that might contain both formative and reflective items. That is, some of these highly correlated variables might be consequents or antecedents of others. For example, 'information quality' is a component of Barnes and Vidgen's (2002) e-Qual instrument, and has items that include accuracy, timeliness and the appropriateness of the presentation format. Since it is possible for information to be accurate but not timely or appropriately presented, and so on, there cannot be a single common cause (information quality) underlying all these measures. There are almost certainly relationships within and between these measures that have not been investigated because of the mistaken assumption that a common-cause factor has been 'discovered'. Because they are not necessarily correlated (remember that a single counter-example suffices), the identified correlations based on which the EFA techniques constructed the common factor must be considered incidental or spurious, albeit typical. It is this typicality that needs to be investigated, as there could be a more distal common cause.

We note that the same issues do not apply to more rigorous techniques such as confirmatory factor analysis (CFA), which specifies a model *a priori* and evaluates the degree to which the data fit the hypothesised model. In this case, the argument for the existence of the factors is external to the analysis (in the theory used to construct the model). Here too, however, the researchers must be cautious to distinguish between proximal and distal factors: latent variables should represent proximal factors (common causes that are most immediately antecedent to the measurement observations) (Hayduk, 1996). In a CFA setting, however, the true model might not necessarily be a simple common-cause latent variable, despite such a model fitting the observed data. This could be due again to incidental but typical correlations that hint at a more distal common cause or at more complex processes underlying the data.

Overall, we suggest that exploratory factor analysis has a limited role to play in building theory of attitudes and behaviour towards technology. It might be preferable to use qualitative methods to explore the perceptions, attitudes and behaviours of users of an information technology in context. The resultant theory can then be specified in advance and evaluated using theory testing methods, such as confirmatory factor analysis and structural equation modelling.

Lack of Replicatory Studies

Description

Replication has been described as the key to generalisation (Hubbard and Armstrong, 1994) and the most important criterion of scientific knowledge (Rosenthal and Rosnow, 1984); so a discussion of replication is related to our earlier discussion of the danger of excessive claims of generalisability. Many widely cited theories in IS are based on samples of several hundred responses, across perhaps three or four organisations. On the basis of this, the authors claim results that are generalisable to comparable contexts, or some might say, over-claim (Seddon and Lyytinen, 2008).

Information systems researchers typically describe their method in considerable detail with the intention that it should be replicable; however, very few replicatory studies in IS are ever conducted or published. Within the social sciences, this problem is not restricted to IS; researchers in marketing (Hubbard and Armstrong, 1994), psychology (Schmidt, 2009) and economics (Arulampalam et al., 1997) have lamented the lack of replicatory studies and the difficulties in getting them published. Replication studies are often considered uninteresting, unoriginal and potentially career limiting (Hubbard and Armstrong, 1994; Schmidt, 2009).

Obviously, a direct replication of a survey conducted in a sample drawn from a population in a social context at a point in time is very difficult; however, there are a number of types of replication and reanalysis studies. These can involve keeping various combinations of the data, the model, the method, the sample, the context or the time horizon of the original study constant, while varying other characteristics (Arulampalam et al., 1997; Schmidt, 2009). Various types of replication can be used for different purposes, including controlling for error or fraud, generalising the results to a larger or different population or verifying the underlying hypothesis by using a different operationalisation (Schmidt, 2009). Given the relatively restricted samples and contexts in which many of our studies are conducted, we are particularly interested in the opportunities for replicating the models and methods from existing IS studies with new populations in different contexts, although other types of replication could also be valuable in some circumstances.

Why Is This A Problem, and What Should We Do Instead?

A review of 20 replicatory studies in marketing found that only 15 per cent of those published confirmed the original findings (Hubbard and Armstrong, 1994). There is no reason to believe the situation would be markedly different in information systems. More reanalysis and replication studies are required to address the potential issue of excessive claims of generalisability, and to ensure confidence in our core body of knowledge.

Although direct replications are difficult outside a laboratory setting, the main obstacle to conducting other types of replicatory studies in social science fields seems to be obtaining raw data from previously published studies for reanalysis or meta-analysis, and the perceived difficulty in publishing the studies (Habing, 2003; Hubbard and Armstrong, 1994; Schmidt, 2009). A variety of solutions has been proposed, many of which have potential relevance to IS researchers.

Journal editors could require that raw data and copies of the procedures used in the research be retained and made available for subsequent reanalysis or meta-analysis. This policy is endorsed by major journals in psychology and marketing (Hubbard and Armstrong, 1994) and could easily be adopted within the IS community. Further, a better understanding of replication and reanalysis, backed by taxonomies such as that provided for psychology research by Schmidt (2009) could be developed. Editorial guidelines for conducting and publishing replications, such as those proposed by Arulampalam et al. (1997) for labour economics could be developed. Journals could have replication sections, special issues or a replication editor (Arulampalam et al., 1997; Hubbard and Armstrong, 1994).

Failure to Consider Alternative Models

Description

Popular data analysis techniques allow the researcher to assess, holistically, an entire theory by means of latent variable models. For covariance analysis, the researcher specifies an *a-priori* model that is tested against the data. What is often neglected is the fact that even if the model fits the data, it might not be the only one. In a trivial example, there are covariance equivalent models that differ significantly (Hayduk, 1996), while in non-trivial cases there might be models that, while not being covariance equivalent, also exhibit good fit to the data. For partial least squares (PLS) analysis, the researcher also specifies an *a-priori* model, but in this case, data are only fitted to the model, with no statistical test being conducted. The objective of the fitting is to maximise the explained variance. Here, it is even more important to recognise that other models could yield as good or better objective fit functions than the one specified. This problem is not unique to survey work using latent variable models. In many experimental studies, researchers do not develop alternative hypotheses to rule out alternative theories or explanations but in fact seek only to confirm their single *a-priori* hypothesis. While traditionally the experimental literature has placed special emphasis on internal validity (ruling out alternative explanations), many IS studies merely pay lip-service to this issue, suggesting that by virtue of an experimental study this point becomes a non-issue. It is, however, frequently their theoretical development leading up to the hypotheses that is at fault. For many studies, the same hypotheses could plausibly follow from alternative theories or alternative chains of reasoning. Thus, while the actual experimental study does in fact rule out so-called nuisance factors, the real issue is alternative theories generating the same hypothesis, and this is typically not sufficiently addressed.

Why Is This A Problem, and What Should We Do Instead?

Given the absence of model tests in PLS and the fact that model tests in structural equation modelling (SEM) can only reject models, these techniques are more suitable to assess the relative performance of different models and theories, rather than the absolute correctness of any particular model. This suggests that researchers should focus their attention on developing either competing models or, for experimental research, competing hypotheses that can differentiate between different models.

Problems with Popper: An inconvenient truth?

Description

In the IS area, but not only in the IS area, many researchers would classify themselves as positivist or at least positivistically inclined but with a more nuanced ontological and epistemological commitment. Despite this, however, many of us fail to heed Popper's argument that we can but disprove theories. Most reviewers and editors will not accept 'negative results' for publication, irrespective of the rigour of the study or the statistical power of the tests.

In an applied field such as ours, 'negative results' might have limited applicability to practice and thus are not as interesting. For example, it might not be very interesting for a business to find that web sites that offer functionality X are no more frequently used than those that do not. We understand that practitioners look for actionable differentiating factors (although a great amount of money could probably be saved if they also looked for non-significant effects).

We believe, however, that a distinction should be drawn between negative observations and negative theories. We agree that for the phenomena themselves, it is important that significant effects be shown; however, we believe that there is a difference between examining a non-significant phenomenon and showing that a particular theory does not explain a significant phenomenon. We can accommodate Popper and still focus on 'positive results'. Related to this is the issue of theoretical development. By not publishing 'positive results' with 'negative explanation' and insisting on rigorous *a-priori* theoretical development, we preclude any serendipitous discovery of theory.

Why Is This A Problem, and What Should We Do Instead?

To many, our theories seem obvious and intuitive. We would like to label this the 'Duh' phenomenon. A charitable interpretation is that all our papers are so well written that, in hindsight, the theory is indeed obvious. A more critical examination suggests that without unexplained data (for example, from a 'significant effect but negative theory' study), or a purely exploratory study, researchers have nothing to go on for developing theory than their own intuition, so that theories are necessarily obvious, intuitive and thus limited. If physicists had limited themselves to theories that their own intuitions alone create, we would probably not today have quantum theory, relativity, superconductivity and a host of others that are counterintuitive. These theories were invented to explain previously unexplained data. In the absence of such data and mysteries, our theories will remain applied commonsense.

In this, we are not calling for exploratory studies completely removed from theory; however, the publication of a significant effect, even when the theory fails to explain it, has potential for the development of new theories. Such effects might be simple ANOVA results or more elaborate in the form of covariance matrices from survey studies. In fact, the *Journal of the AIS* has a policy on providing covariance matrices to just this effect, so that future work can build on the data collected and provide better explanations. We believe this practice should be more widespread. This in no way diminishes the original paper, which should be judged based on the then existing criteria of rigour, originality and relevance.

Summary: Obstacles and alternatives

This chapter offers a brief overview and an initial discussion of a number of issues in the way we review previous research, build and test theory, and develop research designs. While we do not want to contribute to an overly negative anxiety discourse, or suggest that there is not a large amount of valuable, high-quality research published in the IS field, there do appear to be some common threads in the issues we have identified.

One is the unconscious way that some of our research practices are accepted and adopted without critical reflection. This affects the way we present previous literature and tends to result in rather elastic definitions and operationalisations as we attempt to accumulate studies that are in fact incommensurable. It is possible that our insistence on identifying gaps in previous knowledge (perhaps in combination with tenure-track publishing expectations) leads to 'safe' and conservative theorising, and a tendency to add another brick to the theory wall. Another possible issue is the tendency to value some characteristics of theory that are also valued in the natural sciences, such as parsimony and generalisability, without the additional support provided by replicatory studies, reanalysis and meta-analysis. Finally, sometimes knowledge from reference disciplines gets lost in translation, which can result in inappropriate application of theories or methods from important reference disciplines, time lags in applying advances from reference disciplines and a tendency not to apply this knowledge retrospectively. A number of these issues are related, and the list is undoubtedly not comprehensive. A summary is provided in Table 2.1.

Table 2.1 Summary of Obstacles to Theory Building

The literature review	Recommendations
The unconsciously subjective literature review	Use other literature review techniques, including descriptive literature reviews, vote counting and quantitative meta-analysis. Clearly articulate criteria for selection and evaluation of articles. Apply current research standards and heuristics to previously published articles. Recommended reading: Guzzo et al. (1987); King and He (2005).
Issues with meaning variance	Consider a critical analysis or a presentation of the major threads of discourse focusing on the margins and areas of disagreement. Recommended reading: Cooper (1988); Killduff and Mehra (1997); Rosenau (1992); Sylvester and Tate (2008); Sylvester et al. (2007).
Using literature as Lego blocks	Examine what is left unexplained by existing theories. Entertain the possibility of re-conceptualisations of established research areas. Recommended reading: Depends on the nature of the theory.
Reference disciplines lost in translation	Do not rely exclusively on the most recent 'IS translation' of theories from reference disciplines, but go back to original sources as required. Follow current developments in important reference disciplines to avoid time lag in crossing disciplinary boundaries, and to provide criteria for critical evaluation of the methods used in previously published studies. Continue to follow developments in the reference discipline that occur after the theory has been appropriated into information systems. Recommended reading: Depends on the nature of the theory. Some suggestions for theory development based on psychometrics include: Borsboom (2005); Borsboom et al. (2003); Chin et al. (1988); Diamontopoulos (2006); Diamontopoulos and Winklhofer (2001); Edwards and Bagozzi (2000); Jarvis et al. (2003); Petter et al. (2007); Tate and Evermann (2009a).
Theory building and testing	**Recommendations**
Valuing parsimony over accuracy	Create richer, more complex and more nuanced theory that more closely represents the complexity of social contexts and is more likely to be actionable. Recommended reading: Benbasat and Barki (2007); Blalock (1982); Kuhn (1983); Robey and Markus (1998).
Excessive claims of generalisability	Extend theory to include salient design features and affordances of technologies. Conduct more replicatory and reanalysis studies (see below). Recommended reading: Benbasat and Barki (2007); Dubin (1978); Hirschheim (2007).

Failure to distinguish between beliefs, attitudes and behaviours (core theory from social psychology)	Separation of beliefs about technology from the resulting attitudes allows us to accurately model the process by which people form beliefs, attitudes, intentions and behaviours towards technology and address the call to relate our theories to the design features of IS artefacts. The resulting models have the potential to be highly salient to practice and to advance the core of the discipline. Recommended reading: Fishbein and Ajzen (1975); Tate and Evermann (2009b).
Research design	**Recommendations**
Inappropriate use of exploratory factor analysis	Use other methods (for example, qualitative methods) to 'discover' psychological constructs. Hypothesise them a priori and evaluate them using theory testing methods such as CFA or SEMs. Recommended reading: David (1996); Hayduk (1996); McDonald (1996); Maraun (1996a, 1996b, 1996c); Mulaik (1996); Rozeboom (1996); Schonemann (1996a, 1996b); Steiger (1996a, 1996b); the SEMNET online forum.
Lack of replicatory studies	Develop methodological and editorial guidelines for conducting and publishing replication and reanalysis studies. Recommended reading: Arulampalam et al. (1997); Hubbard and Armstrong (1994); Schmidt (2009).
Failure to consider alternative models	Develop, test and publish competing models. Recommended reading: Hayduk (1996).
Problems with Popper	Publish significant effects, even if the theory does not explain them. Make raw data available for future work to provide alternative explanations for the data collected.

Implications

As IS is a practice-based field, IS researchers tend to have a strong focus on conducting empirical research to explain phenomena with real-world salience. Simultaneously, there is a widely held belief that as IS is an academic discipline, our concern should be with developing generalisable theory. These two areas of focus are sometimes in conflict. A further complication is that, despite the assertions of Baskerville and Myers (2002) that IS can be a reference discipline, many of our theories are adaptations and appropriations of theory from reference disciplines. We have a tendency to extend these IS appropriations without re-engaging deeply with either the original theories or their philosophical underpinnings. This can lead to a rather formulaic approach to theory building that follows 'best practice' in our field uncritically.

The major implication of this chapter is that quality theory building requires a deep, thoughtful and critical engagement by the researcher with the philosophy of science, reference disciplines, methods and data analysis literature, and the assumptions made at all stages of the research process ranging from the selection and presentation of literature, the goals and scope of the theory-building process to the selection of the research design. We recognise this is a big ask, but we believe it is essential to improve the quality of our theory building.

Conclusion

Overall, despite 20 to 30 years of quantitative research in theories of attitudes and behaviours towards information technology there is relatively little agreement about core constructs, definitions and operationalisations. This is not to suggest that all is hopeless, or that there are not significant amounts of high-quality research. There is, however, evidence that there is considerable room for improvement. In this chapter, we identify a number of research issues and practices that could be impeding our disciplinary progress. We also introduce a range of literature from reference disciplines offering more in-depth discussions and alternative perspectives that might stimulate further debate.

The contribution of this chapter is to stimulate debate about some common methodological assumptions and practices of IS researchers, to present contrarian views and alternatives and to showcase recent contributions from reference disciplines.

If the practices we have critiqued are adopted uncritically then we will perpetuate the situation expressed in the meme 'if you do something one hundred times you should not expect a different result on the hundred and first time'. Our earnest hope is that this discussion will promote debate, critical reflection and overall quality improvement for theory building in this critical aspect of IS research.

References

Ajzen, I. (1991). The theory of planned behavior. *Organizational Behavior and Human Decision Processes, 50*(2), 179–211.

Arulampalam, W., Hartog, J., MaCurdy, T., & Theeuwes, J. (1997). Replication and re-analysis. *Labour Economics, 4*(2), 99–105.

Babakus, E., & Boller, G. (1992). An empirical assessment of the SERVQUAL scale. *Journal of Business Research, 24*(3), 253–68.

Ballantine, J., Bonner, M., Levy, M., Martin, A., Munro, I., & Powell, P. L. (1996). The 3-D model of information systems success: the search for the dependent variable continues. *Information Resources Management Journal, 9*(4), 5–14.

Barki, H. (2008). Thar's gold in them thar constructs. *The Database for Advances in Information Systems, 39*(3), 9–20.

Barnes, S., & Vidgen, R. (2002). An integrative approach to the assessment of e-commerce quality. *Journal of Electronic Consumer Research, 3*(3), 114–27.

Baskerville, R. L., & Myers, M. D. (2002). Information systems as a reference discipline. *MIS Quarterly, 26*(1), 1–14.

Benbasat, I., & Barki, H. (2007). Quo vadis, TAM? *Journal of the Association for Information Systems, 8*(4), 211–18.

Benbasat, I., & Zmud, R. W. (2003). The identity crisis within the IS discipline: defining and communicating the discipline's core properties. *MIS Quarterly, 27*(2), 183–94.

Bhattacherjee, A., & Premkumar, G. (2004). Understanding changes in belief and attitude toward information technology usage: a theoretical model and longitudinal test. *MIS Quarterly, 28*(2), 229–54.

Blalock, H. (1982). *Conceptualisation and Measurement in the Social Sciences.* Beverley Hills, Calif.: Sage.

Borsboom, D. (2005). *Measuring the Mind: Conceptual issues in contemporary psychometrics.* Cambridge, UK: Cambridge University Press.

Borsboom, D., Mellenbergh, G. J., & van Heerden, J. (2003). The theoretical status of latent variables. *Psychological Review, 110*(2), 203–19.

Carman, J. (1990). Consumer perceptions of service quality: an assessment of the SERVQUAL dimensions. *Journal of Retailing, 66*(1), 33–55.

Chin, J., Diehl, V., & Norman, K. (1988). Development of an instrument measuring user satisfaction of the human–computer interface. *Proceedings of the SIGCHI Conference on Human Factors in Computing Systems*, Washington, DC, 213–18.

Cooper, H. M. (1988). Organising knowledge syntheses: a taxonomy of literature reviews. *Knowledge, Technology and Policy, 1*(1), 104–26.

Cooper, J. (2007). *Cognitive Dissonance: 50 years of a classic theory*. London, UK: Sage Publications.

Curd, M., & Cover, J. A. (1998). *Philosophy of Science: The central issues*. New York, NY: W. W. Norton & Company.

David, B. (1996). Comment on: metaphor taken as math: inderterminancy in the factor model. *Multivariate Behavioral Research, 31*(4), 551–4.

Davis, F., Bagozzi, R., & Warshaw, P. (1989). User acceptance of computer technology: a comparison of two theoretical models. *Management Science, 35*(8), 982–1003.

DeLone, W. H., & McLean, E. R. (1992). Information systems success: the quest for the dependent variable. *Information Systems Research, 3*(1), 60–95.

Diamontopoulos, A. (2006). The error term in formative measurement models: interpretation and modelling implications. *Journal of Modelling in Management, 1*(1), 7–17.

Diamontopoulos, A., & Winklhofer, H. (2001). Index construction with formative indicators: an alternative to scale development. *Journal of Marketing Research, 38*(2), 269–77.

Dubin, R. (1978). *Theory Building*. New York, NY: Free Press.

Edwards, J. R., & Bagozzi, R. P. (2000). On the nature and direction of relationships between constructs and measures. *Psychological Methods, 5*(2), 155–74.

Evermann, J. (2005a). Organizational paradigms and organizational modelling. *Proceedings of the Workshop on Business and IT Alignment (BUSITAL)*, Porto, Portugal, 230–9.

Evermann, J. (2005b). Towards a cognitive foundation for knowledge representation. *Information Systems Journal, 15*(2), 147–78.

Evermann, J., & Tate, M. (2009a). Building theory from quantitative studies, or, how to fit SEM models. *Proceedings of the 30th International Conference on Information Systems*, Phoenix, Ariz.

Evermann, J., & Tate, M. (2009b). Constructs in the mist: the lost world of the IT artifact. *Proceedings of the 30th International Conference on Information Systems*, Phoenix, Ariz.

Festinger, L. (1957). *A Theory of Cognitive Dissonance*. Stanford, Calif.: Stanford University Press.

Fishbein, M., & Ajzen, I. (1975). *Belief, Attitude, Intention and Behaviour: An introduction to theory and research*. Reading, Mass.: Addison-Wesley.

Goodhue, D. L., & Thompson, R. L. (1995). Task-technology fit and individual performance. *MIS Quarterly, 19*(2), 213–36.

Guzzo, R. A., Jackson, S. E., & Katzell, R. A. (1987). Meta-analysis analysis. *Research in Organisational Behavior, 9*, 407–22.

Habing, B. (2003). *Exploratory factor analysis*. Working Paper. University of South Carolina, Columbia, SC.

Hayduk, L. (1996). *LISREL: Issues, debates and strategies*. Baltimore, Md: The Johns Hopkins Press Ltd.

Hirschheim, R. (2007). Introduction to the special issue on 'Quo Vadis TAM: Issues and reflections on technology acceptance research'. *Journal of the Association for Information Systems, 8*(7), 203–5.

Hirschheim, R., & Newman, M. (1991). Symbolism and information systems development: myth, metaphor, and magic. *Information Systems Research, 2*(1), 29–62.

Hubbard, R., & Armstrong, J. S. (1994). Replications and extensions in marketing: rarely published but quite contrary. *International Journal of Research in Marketing, 11*(4), 233–48.

Jarvis, C. B., MacKenzie, S., & Podsakoff, P. M. (2003). A critical review of construct indicators and measurement model misspecification in marketing and consumer research. *Journal of Consumer Research, 30*(2), 199–218.

Keen, P. (1980). MIS research: reference disciplines and a cumulative tradition. *Proceedings of the 1st International Conference on Information Systems*, Philadelphia, Pa.

Kendall, J., & Kendall, K. (1993). Metaphors and methodologies: living beyond the systems machine. *MIS Quarterly, 17*(2), 149–71.

Killduff, M., & Mehra, A. (1997). Postmodernism and organisational research. *The Academy of Management Review, 22*(2), 453–81.

King, W. R., & He, J. (2005). Understanding the role and methods of meta-analysis in IS research. *Communications of the Association for Information Systems, 16*(1), 665–86.

Kling, R., & Iacono, S. (1984). The control of information systems developments after implementation. *Communications of the ACM, 27*(12), 1218–26.

Kuhn, T. S. (1983). Rationality and theory choice. *Journal of Philosophy, 80*(10), 563–70.

Lyytinen, K., & King, J. L. (2004). Nothing at the centre? Academic legitimacy in the information systems field. *Journal of the Association for Information Systems, 5*(6), 220–64.

McDonald, R. (1996). Consensus emerges: a matter of interpretation. *Multivariate Behavioral Research, 31*(4), 663–72.

Maraun, M. (1996a). Meaning and mythology in the factor analysis model. *Multivariate Behavioral Research, 31*(4), 603–16.

Maraun, M. (1996b). Metaphor taken as math: indeterminacy in the factor analysis model. *Multivariate Behavioral Research, 31*(4), 517–38.

Maraun, M. (1996c). The claims of factor analysis. *Multivariate Behavioral Research, 31*(4), 673–89.

Milton, S., & Kazmierczak, E. (2004). An ontology of data modelling languages: a study using a common-sense realistic ontology. *Journal of Database Management, 15*(2), 19–38.

Mulaik, S. (1996). On Maraun's deconstructing of factor indeterminancy with constructed factors. *Multivariate Behavioral Research, 31*(4), 579–92.

Noy, N., & Hafner, C. (1997). The state of the art in ontology design: a survey and comparative review. *AI Magazine, 18*(3), 53–74.

Parasuraman, A., Zeithaml, V., & Berry, L. (1988). SERVQUAL: a multi-item scale for measuring consumer perceptions of service quality. *Journal of Retailing, 64*(1), 12–40.

Parasuraman, A., Zeithaml, V., & Berry, L. (1991). Refinement and reassessment of the SERVQUAL scale. *Journal of Retailing, 67*(4), 420–51.

Pavlou, P. A., & Gefen, D. (2005). Psychological construct violations in online marketplaces: antecedents, consequences and moderating role. *Information Systems Research, 16*(4), 372–99.

Petter, S., DeLone, W., & McLean, E. (2008). Measuring information systems success: models, dimensions, measures and interrelationships. *European Journal of Information Systems, 17*(3), 236–63.

Petter, S., Straub, D., & Rai, A. (2007). Specifying formative constructs in information systems research. *MIS Quarterly, 31*(4), 623–56.

Pitt, L., Watson, R., & Kavan, B. (1995). Service quality: a measure of information systems effectiveness. *MIS Quarterly, 19*(2), 173–87.

Pittayachawan, S., Singh, M., & Corbitt, B. (2008). A multitheoretical approach for solving trust problems in B2C e-commerce. *International Journal of Networking and Virtual Organisations, 5*(3–4), 369–95.

Reed, M. (1992). *The Sociology of Organisations: Themes, perspectives, and prospects*. Herfortdshire, UK: Harvester Wheatsheaf.

Robey, R., & Markus, M. L. (1998). Beyond rigor and relevance: producing consumable research about information systems. *Information Resources Management Journal, 11*(1), 7–15.

Rosenau, P. M. (1992). *Postmodernism and the Social Sciences: Insights, inroads, and intrusions*. Princeton, NJ: Princeton University Press.

Rosenthal, R., & Rosnow, R. L. (1984). *Essentials of Behavioral Research: Methods and data analysis*. New York, NY: McGraw-Hill.

Rozeboom, W. (1996). What might common factors be? *Multivariate Behavioral Research, 31*(4), 555–70.

Schmidt, S. (2009). Shall we really do it again? The powerful concept of replication is neglected in the social sciences. *Review of General Psychology, 13*(2), 90–100.

Schonemann, P. H. (1996a). Syllogisms of factor indeterminacy. *Multivariate Behavioral Research, 31*(4), 651–4.

Schonemann, P. H. (1996b). The psychopathology of factor indeterminancy. *Multivariate Behavioral Research, 31*(4), 571–7.

Seddon, P. (2006). MISQ new author's workshop. Paper presented at the 11th Pacific-Asia Conference on Information Systems (PACIS), Auckland, New Zealand, July.

Seddon, P., & Lyytinen, K. (2008). Panel on generalisability in information systems theory. Paper presented at the 19th Australasian Conference on Information Systems (ACIS), Christchurch, New Zealand, December.

Steiger, J. H. (1996a). Coming full circle in the history of factor indeterminacy. *Multivariate Behavioral Research, 31*(4), 617–30.

Steiger, J. H. (1996b). Dispelling some myths about factor indeterminacy. *Multivariate Behavioral Research, 31*(4), 539–50.

Sum, C.-C., Lee, Y.-S., Hays, J., & Hill, A. (2002). Modeling the effects of a service guarantee on perceived service quality using alternating conditional expectations (ACE). *Decision Sciences, 33*(3), 347–83.

Sureshchandar, G. S., Rajendran, C., & Anantharaman, R. N. (2002). Determinants of customer-perceived service quality: a confirmatory factor analysis approach. *Journal of Services Marketing, 16*(1), 9–34.

Sylvester, A., & Tate, M. (2008). Beyond the 'mythical centre': an affirmative post-modern view of SERVQUAL research in information systems. Paper presented at the 16th European Conference on Information Systems (ECIS), Galway, Ireland, June.

Sylvester, A., Tate, M., & Johnstone, D. (2007). *Re-presenting the literature review: a rich picture of service quality research in information systems*. Paper presented at the 11th Pacific Asia Conference on Information Systems (PACIS), Auckland, New Zealand, July.

Tannenbaum, A. S. (1962). Control in organizations: individual adjustment and organizational performance. *Administrative Science Quarterly, 7*(2), 236–57.

Tate, M., & Evermann, J. (2009a). Descendents of ServQual in online services research: the end of the line? Paper presented at the 15th Americas Conference on Information Systems (AMCIS), San Francisco, Calif., August.

Tate, M., & Evermann, J. (2009b). Perceptive users with attitudes: some heuristics of theorizing. *Proceedings of the International Conference on Information Systems (ICIS 2009)*, Phoenix, Ariz.

Teo, T., Srivastava, S., & Jiang, L. (2009). Trust and electronic government success: an empirical study. *Journal of Management Information Systems, 25*(3), 99–131.

Uschold, M., & Gruninger, M. (1996). Ontologies: principles, methods, and applications. *Knowledge Engineering Review, 11*(2), 93–136.

Venkatesh, V., Morris, M., Davis, G. B., & Davis, F. D. (2003). User acceptance of information technology: towards a unified view. *MIS Quarterly, 27*(3), 425–78.

vom Brocke, J., Simons, A., Niehaves, B., Riemer, K., Plattfaut, R., & Cleven, A. (2009). Reconstructing the giant: on the importance of rigour in documenting the literature search process. Paper presented at the 17th European Conference on Information Systems (ECIS), Verona, Italy, June.

Walsham, G. (1991). Organizational metaphors and information systems research. *European Journal of Information Systems, 1*(12), 83–93.

Wand, Y., & Weber, R. (1993). On the ontological expressiveness of information systems analysis and design grammars. *Journal of Information Systems, 3*(4), 217–37.

Weber, R. (1997). *Ontological Foundations of Information Systems.* Melbourne, Vic.: Coopers & Lybrand.

Webster, J., & Watson, R. T. (2002). Analyzing the past to prepare for the future: writing a literature review. *MIS Quarterly, 26*(2), xiii–xxiii.

Wyssusek, B. (2005). On the foundation of the ontological foundation of conceptual modeling grammars: the construction of the Bunge–Wand–Weber ontology. *Proceedings of the 1st International Workshop on Philosophical Foundations of Information Systems Engineering (PHISE'05)*, Porto, Portugal.

Yang, Z., Cai, S., Zhou, Z., & Zhou, N. (2005). Development and validation of an instrument to measure user-perceived service quality of information presenting web portals. *Information and Management, 42*(4), 575–89.

3. Untangling Causality in Design Science Theorising

Dirk S. Hovorka
Bond University

Shirley Gregor
The Australian National University

Abstract

Although design science research aims to create new knowledge through design and evaluation of artefacts, the causal agency through which artefacts obtain predicted outcomes is frequently under-specified. Within this domain of knowledge, six types of causal reasoning can be applied by researchers to more clearly articulate why desired outcomes will result from the implementation of the artefact. In addition, reflecting on the causal foundations of the design will enable more definitive evaluation of the design theory and scientific explanation of the behaviour of the artefact-in-use. The framework proposed here is based on an extensive literature in causal theory and the implications are that researchers will be able to articulate the causal reasoning used in design science theorising.

Introduction

Design science research (DSR) seeks to create new knowledge through the process of designing, building and evaluating information system artefacts. Designed systems are teleological in nature: they have an intended purpose, and designers and users have expectations of specific observable outcomes as a direct result of implementation and use. The purpose of design lies in shaping artefacts and events to create a more desirable future (Boland, 2002). As these systems are intended to mediate or intervene in personal, group or organisational activities to produce specific outcomes, they are perceived to have causal agency, either implicitly or explicitly. Any proposed design solution prescribes technological rules that provide general instructions for building an artefact intended to produce a specific outcome (Bunge, 1967; van Aken, 2004).

In addition, these technological rules can form part of a design theory that links specific architectures and outcomes and, additionally, predicts and explains the outcomes that obtain from those structures. The suggestion that DSR is intended to contribute to our theoretical knowledge (Gregor, 2006; Gregor and Jones, 2007; Venable, 2006) has become more generally accepted. Beginning at least with Aristotle, to know the causes of things was fundamental to the explanatory disciplines and is still characteristic of modern science (Bunge, 2008; Salmon, 1998). Thus, reasoning about causality is required by both the designers of artefacts in their construction activities and development of design theory and the researchers who study the behaviour and effects of artefacts-in-use for evaluation purposes, to inform future design, and in building theory about designed artefacts.

Despite the implicit reliance on causal reasoning and its centrality in theory building, the problem of causality in the DSR literature has been little addressed or has been addressed in a relatively simplistic fashion. Only rarely are causal connections explicitly specified in DSR and, when identified, such connections are only very generally described. In many cases where kernel theories are specified, researchers retreat behind the simplification of cause expressed in statements that a specific kernel theory provides justification for the prediction and explanation of the desired outcomes.[1] Direct connections between the causal mechanisms of the kernel theory and how these causal mechanisms will be instantiated in the design to produce the expected outcome are rarely described. Work that studies artefacts-in-use frequently employs statistical methods, where questions of causality are avoided or glossed over. Although, as Venable (2006) points out, it is possible to create a design that successfully produces a better state of affairs in the problem space, but not know how or why it works. But this limitation constrains the contribution to the knowledge base and our ability to apply that knowledge in other domains.

The extensive discussion of causality in the philosophy of science literature precludes anything but a modest review in a chapter such as this. We draw on this literature, however, to provide a basic conceptualisation of causation and to propose a framework that might guide researchers in the process of design theorising and in the evaluation of artefacts and the knowledge discovered as a result of artefact construction and use. The goal of the framework is to sensitise researchers to causal reasoning in design science research. The framework itself is not prescriptive but can be used to identify and refine causal reasoning as

1 A kernel theory, or 'justificatory knowledge' or 'micro-theory', is the explanatory knowledge that links other components of a design theory: design goals, principles, processes and materials. It is the knowledge that explains 'why' a particular design is expected to work and thus involves causal reasoning (Gregor and Jones, 2007). For example, knowledge from cognitive science explains why certain interface design principles are valid. Kernel theory can come from reference disciplines such as cognitive science or from other design theory (for example, theory from aesthetics).

it is applied to research in the interior prescriptive mode (the construction of artefacts) and also in the exterior descriptive mode (use of artefacts in socio-technical systems) (Gregor, 2009). More explicit causal reasoning will enable design science (DS) researchers to better select and apply kernel theories, to evaluate design principles more deeply and to provide stronger knowledge claims when evaluating socio-technical systems. It will also inform descriptive research of artefacts-in-use, which will aid the development of theory that better informs design science research.

Importance of Causal Thinking in Design Science 'Theorising'

Design is inherently based on causal claims or assumptions. All human activities intended to shape or control future events are based on causal inference and the design is a 'specifications of actions to be taken (often in a specified sequence) to achieve the intended consequence' (Argyris, 1996, p. 396). In theorising about design of information systems (IS), the causal agency may be transferred to the technology and/or to the users' interactions with the technology. In some cases, the causal agency may exist at the level of organisational actions in the implementation or control of the technology.

The IS community has engaged in considerable discussion and argument about the nature and relevance of theory in DSR (Gregor and Jones, 2007; Hooker, 2004; Venable, 2006) and we do not seek to enter this debate in this essay. But whether a researcher produces a set of design principles, a design theory or a set of technological rules, the design of a teleological artefact contains warrants about antecedent–consequent relationships that must be grounded in existing knowledge. Goldkuhl (2004) identified empirical, theoretical and internal types of grounding. Of particular importance in this discussion is the explanatory aspect of the theoretical grounding type that is found in the reliance on kernel theories as a basis for design.

Selection of kernel theories is vexing because in most problem domains multiple, sometimes contradictory, theories exist. The design researcher must select, from the possible theory base, kernels that seem relevant to the design problem often without full knowledge of their suitability to the design problem. As an external source of theoretical grounding, examination of the causal claims in the kernel theories potentially provides a stronger grounding for the resultant design theory. In addition, explicit recognition of the causal commitments assumed in the design become clear research questions for the evaluation phase and might lead to improved knowledge contributions from the DSR process.

Causal reasoning will also enable a better contextualisation by identifying how, when and where kernel theories are applicable and what interactions between kernel theories can be expected.

We note that not all design theorising is based on kernel theories and, even when present, kernel theories might serve as creative inspiration rather than a source of logical derivation of design theory (Goldkuhl, 2004). Causal reasoning requires the researcher to evaluate in what context, and for which system, participants and tasks each specific kernel theory was or was not relevant to the design (Hovorka and Germonprez, 2009). In these instances, the increased ability for appropriate evaluative and knowledge construction is a sufficient argument for description of the *assumed* causal mechanisms.

Background

As a research approach, DSR is based on the idea that knowledge of the world can be obtained through a 'problem identification-build-evaluate-theorise' process (Hevner et al., 2004). In the problem identification phase, researchers identify a problem domain in which the intervention of a new artefact into an environment will produce a change in specified outcomes. This phase inherently involves causal thinking—specific design characteristics of the artefact have causal agency to produce outcomes that will solve the identified problem. Causal reasoning also appears during theory development. As noted by Gregor (2006), a theory that explains, predicts or prescribes is offering a causal explanation by identifying what antecedent conditions will result in specific consequents.

The identification of causality is, however, extremely problematic, and among philosophers the very concept of causation has suffered numerous deaths, including strong critiques resulting from quantum theory and logical positivism (Bunge, 2008). Yet reasoning about causation—whether conceived as complete determinism (that is, Laplace's daemon), as a specific relationship between entities (Salmon, 1998) or as regularities in perception leading to cognitive beliefs (Hume, 2004)—is an instinctive tendency of human behaviour (Bunge, 2008). Humans are curious, and the case can be made that their survival depends on determining why events occur and how to intervene to shape their environments in a desirable way. Yet arguably, design of the artefacts that are the object of study for DSR is founded on many different types of determinism and is lumped into a causal language that distorts the real contributions to knowledge. Although the word 'cause' is often omitted in research papers, our ideas of determination of effects based upon designed artefacts bear the stamp of causality (Bunge, 2008). Untangling and clarifying precisely how design theory and artefacts determine outcomes will benefit both our design theories and the knowledge created by design science research.

Much of the analytic thinking about causation is based on the assumption that individual causes are independent of each other—that changes in one factor will affect the outcome but will not change any other factor. Yet this assumption does not hold in the real-world situations in which DSR operates. Moreover, the model of the world assumed in DSR is one of general linear reality (Abbott, 2001), in which the order of events does not influence the outcomes. Yet the outcomes resulting from the artefacts in use vary with time and context, the order in which information is presented affects human decisions and causal agency changes for different stakeholders over time.

The concept of causality has a long history that can be traced back to Aristotle and the early Greek philosophers, who recognised a fundamental distinction between descriptive knowledge saying that something occurred and explanatory knowledge saying why something occurred. Notably, Aristotle's doctrine identified four causes (*aitia*) (from a translation by Hooker, 1996):

- *material cause*: 'that out of which a thing comes to be, and which persists' (that is, what a thing is made of)
- *formal cause*: 'the statement of essence' (that is, the form and pattern that define something as 'this' rather than 'that')
- *efficient cause*: 'the primary source of change' (that is, the designer or maker of something)
- *final cause*: 'the end (*telos*), that for the sake of which a thing is done' (for example, health is the cause of exercise).

Modern science has focused primarily on efficient causes, the agents that bring about change, with material causes assumed to be that which is changed.

There are numerous means of reasoning about causality and at different times and contexts we might ascribe causality through different theoretical conceptions. Practical criteria for determination of causality were presented by J. S. Mill (1882) as: 1) the cause has to precede the effect in time; 2) the cause and effect must be related; and 3) other explanations of the cause–effect relationship have to be eliminated (Shadish et al., 2002). These criteria are still relevant, but they are overly simplistic when dealing with the construction of information technology-based artefacts.

For the purpose of an analytic framework, we begin with consideration of two important views of causation: *event causation* and a form of *agent causation* (largely following Kim, 1999).

Event causation is the causation of an event by some other event or events. A computer virus or power outage will cause system disruption, but the opposite is not true. Kim (1999) distinguishes four prominent approaches to analysing event causality.

1. *Regularity analysis* (constant conjunction or nomological analysis): This is the type of causality common in the natural sciences and is based on uniform and constant covering laws. 'There are some causes, which are entirely uniform and constant in producing a particular effect; and no instance has ever been found of any failure or irregularity in their operation' (Hume, 2004, p. 206). It is argued, however, that due to the complexity and variability of human behaviour, this type of regularity should not be expected or sought in the social sciences (Fay, 1996; Little, 1999).
2. *Counterfactual analysis:* This means of analysis posits that what qualifies an intervention as a cause is the fact that if the intervention had not occurred, the outcome would not have happened (the cause is a necessary condition). To say that striking a drinking glass caused the glass to break is to say that the breaking was counterfactually dependent on the strike. If the glass had not been struck it would not have broken (*ceteris paribus*) (Collins et al., 2004).
3. *Probabilistic analysis:* This type of causality was recognised by Hume (2004, p. 206)—in comparison with universal laws, 'there are other causes, which have been found more irregular and uncertain; nor has rhubarb always proved a purge, or opium a soporific to everyone, who has taken these medicines'. This view of causal analysis is thought to be suited to the social sciences, where the lack of a closed system and the effects of many extraneous influences make other causal analysis difficult to undertake. 'To say that C is the cause of E is to assert that the occurrence of C, in the context of social processes and mechanisms F, brought about E, or increased the likelihood of E' (Little, 1999, p. 705).
4. *Manipulation analysis:* This conception of causation entails the idea that an intervention in a system will influence the outcomes. That is, the cause is an event (an act) that we can manipulate or perform to bring about an effect—for example, pressing a switch turns a light off. This practically oriented conception can identify knowledge useful for specific kinds of prediction problems. It contains elements of variance such that probabilistic effects can be accounted for. More importantly, it provides a separate inferential step that allows us to differentiate the case where two variables are correlated, from the case where it is claimed that one variable will respond when under manipulation by the other (Woodward, 2003).

In addition to the above four forms of analysis pertaining to event causation, it is useful to consider for DSR the separate category of agent causation. Agent causation 'refers to the act of an agent (person, object) in bringing about a change' (Kim, 1999, p. 125). Thus, my flicking the light switch is the cause of the light turning on. It can be seen that agent causation analysis in general could be seen as reducible to manipulation event analysis. That is, the movement of my hand (an event) caused the light to come on (another event) and both these events were preceded by other events in a chain (walking through the door, perceiving that the room was dark). In this case, the act of an agent can be seen as *reactive*. It is a consequence of the agent's beliefs, attitudes and environmental inputs (Pearl, 2000).

Some have claimed, however, that one form of agent causation is not reducible to event analysis—namely, *substantival causation* (Kim, 1999). This form of causation is particularly relevant in design disciplines and we will distinguish it with a fifth form of causal analysis in our framework.

5. *Substantival causation analysis* (mental causation): This form of analysis recognises the creation of a novel or genuinely new substance or artefact by a human or humans, going beyond the mere change or manipulation of existing substances or their rearrangement. Many inventions would be examples of the effects of this type of causation—for example, the first telescope, the first bicycle and the first decision support system. The ability of humans to project virtual realities which do not yet exist (Ramiller, 2007) is a necessary but not sufficient cause in the design of artefacts. Recognising this type of causality requires recognition that humans have free will and can choose to do or create things that did not exist before, and these things themselves can play a part in other causal relationships. This type of causation recognises the *deliberative* (rather than reactive) behaviour of humans in exercising free choice (Pearl, 2000). The issue of the connection between the mental deliberations of humans and their consequent observable actions is part of a larger mind–body problem, which is beyond the scope of this essay. We will, however, distinguish this type of causation separately, because of its implications for design work.

Some further discussion of concepts relevant to causality is necessary to clarify some basic assumptions underlying the essay and our usage of terms. First, a cause is seen as an event or action that results in a change of some kind. If nothing changes then there is no cause and no consequent effect—that is, there is no change of state (Schopenhauer, 1974). Further, we have to consider the distinction between active causes and contextual causal conditions (which are more static). These each pertain to the issue of necessary and sufficient conditions, as these are central to many arguments for causality and to counterfactual analysis specifically. A counterfactual argument rests on the claim that effect *E* would not have occurred if cause *C* had not occurred; in this case *C* is a *necessary cause* for *E*. To use a highly simplified example, the application of a burning match to a material could be seen as a necessary cause for a fire to light; however, there are other contextual conditions that are also needed for a material to ignite—for example, there must be enough oxygen present. Thus, though the match is necessary, it is not sufficient to cause a fire in the absence of other contributing contextual factors. But, taken together, the active cause and the causal condition (striking match plus oxygen) could be considered necessary and sufficient conditions for the fire to light. But even in this relatively simple case, there are problems in specifying all of the contextual conditions that are needed for both necessity and sufficiency. It might be that the active causal intervention of the burning match is not necessary, because some other active event could cause the fire to light (for example, lightning, spontaneous combustion, a spark from an electrical wiring fault). Further, it is difficult to specify all the contextual conditions that are necessary; in this case we have not specified that the ignited material must be flammable and it must be dry (for example, there must be an absence of water). The problem of complete determination of necessary and sufficient conditions verges on the impossible except for very simple, well-defined and closed systems.

It is for this reason that the words *ceteris paribus* (all else being equal) are added to claims to narrow the scope of the claim. For example, with our relatively simple case of the fire, we could claim 'given the existing state conditions (flammable material, oxygen, absence of water) and in the absence of other causes (spontaneous combustion, electrical spark), if the match had not been brought into close proximity to the material, the fire would not have started'. The causal claim is that close proximity of a burning match, *ceteris paribus*, is a necessary and sufficient condition for starting a fire in other (similar) situations. Here we see that causal claims are a form of generalisation (Lee and Baskerville, 2003) in which a theory specifying causal relationships is generalisable to other instances in similar contexts. In this way all design theories are claims that, *ceteris paribus*, an artefact built with the specified principles will cause the predicted outcome. Implied but rarely recognised in DSR is the necessary condition that the artefact be implemented and used successfully. As noted by Venable (2006), the problem space is composed of many related and potentially conflicting concepts, goals and stakeholders and the designed artefact will cause the predicted outcomes only if it is used in a manner consistent with the problem space as defined in the meta-requirements.

In socio-technical systems, however, we have to deal with situations where the number of causal conditions is large and there can be considerable uncertainty about the nature of linkages between cause and effect (Fay and Moon, 1996). Problem spaces in which artefacts will be implemented only rarely (if ever) fit *ceteris paribus* conditions. In such situations it is useful to consider probabilistic reasoning about necessary and sufficient conditions. Pearl (2000) has advanced thinking in this area, and provides detailed coverage of how such reasoning can be dealt with for identification of causality using statistics. Pearl (2000, p. 284) shows how the 'probability of necessity' can be thought of in terms such as 'the probability that disease would not have occurred in the absence of exposure' to an infection. The disease might occur in only 1 per cent of cases without exposure. If you are not exposed you have a 99 per cent chance of not getting the disease; exposure is 'almost' a necessary condition. Similarly, the 'probability of sufficiency' can be expressed in terms such as the probability that a healthy, unexposed individual would have contracted the disease had he or she been exposed. The disease might follow exposure in 70 per cent of cases. There are links between this type of reasoning and the type of analysis that is needed in information systems. For example, the probability of necessity for module testing to ensure all errors are detected in programming is 99 per cent (1 per cent of cases would be error free if no module test occurs). The probability of necessity emphasises the absence of alternative causes that are capable of explaining the effect. The probability of sufficiency of a committed project champion is 80 per cent (80 per cent of cases with a committed project champion will be successful). The probability of sufficiency emphasises the presence of active causal processes that can produce the effect. The intricacies of determining necessary and sufficient conditions are laboured somewhat here because of their importance in reasoning in design science research. It is a very common form of analysis even if not recognised explicitly. Examples are cross-case analyses where an attempt is made to identify 'key' factors that are necessary, sufficient or both for some outcome to occur.

Some other aspects of causality are worthy of note for design disciplines. In the case of the fire lighting, a necessary condition is that the fire material is flammable. That is, the fire consumes fuel that is conducive to being lit. In information systems design fields, particularly in human–computer interaction, something like this notion is captured by the idea of 'affordance'. As explained by Norman (1988), the affordances of an object are the action possibilities that are readily perceivable by an actor because of the object's design characteristics. An example is a door that has no handle on the side that is to be pushed rather than pulled.

Another consideration is the causal characteristic of random interplay (Bunge, 2008), which results in emergent and unpredictable effects. Although these effects cannot be controlled or predicted, conditions that enable emergent behaviours and outcomes to arise from the lack of tightly coupled integration of components can be *designed for* in the evolution or secondary design of information systems (Germonprez et al., 2007). Systems in use consistently show unexpected consequences (Dourish, 2006; Winograd and Flores, 1986), and Ciborra (2002, p. 44) notes that new systems of value emerge when users are 'able to recognise, in use, some idiosyncratic features that were ignored, devalued or simply unplanned'.

The concepts of both affordance and tailorable design are important because they are potential *causal conditions* that enable or constrain actions with the artefact that cannot be foreseen at the time of the design. They are potentially players in indeterminate chains of causal events. We will recognise the importance of this type of causality by distinguishing it as a sixth type of causal reasoning in design science research.

6. *Enabling causal condition* analysis: This analysis involves consideration of how artefact characteristics and conditions constrain or enable subsequent causal outcomes during use. The important characteristic is that the inclusion or exclusion of particular design characteristics will change the likelihood of the outcomes. This type of analysis is similar to type four, in which an intervention (an event or act) brings about an effect or makes an effect more likely. Here, however, we are separating active causes and contextual causal (enabling) conditions. In the example we gave previously, the act of striking a match was the active causal condition, whereas the placing of combustible material in the room by an agent was an enabling causal condition. A further example is perceived affordance, in which elements allow or encourage *possible* actions that are latent in the design. Examples include the scroll wheel on a computer mouse and roll-over text that informs users what will happen if they select a specific hyperlink. Another example of an enabling causal condition is the use of component architectures and recognisable conventions (Germonprez et al., 2007) that enable users to recognise conventional functions of component parts that can be reassembled into new patterns or adapted to new task functions. The design principles for the artefact are loosely coupled to the world so that users can create new structural couplings in alignment with their domain of action (Winograd and Flores, 1986). Design conventions such as icons that resemble functions performed (for example, a waste bin for 'delete', an hourglass for 'wait') guide people towards correct usage. This is a probabilistic cause in that most people familiar with icon conventions and symbols will understand the implied function and act accordingly.

The focus in design is often on obtaining a specific set of outcomes. But design can also include the goal of preventing specific outcomes (for example, preventing unauthorised system access, designing a 'rigid' artefact that users cannot modify in use). In this case the designer seeks to identify and eliminate necessary conditions for the undesired outcome or to find causes that obtain conditions that prevent the undesired outcome.

Note that we are discussing causality with reference to the work of a number of scholars who have made important contributions in this area. Our arguments have some congruence with ideas expressed in contemporary 'critical realism'—a philosophical approach based on work by Bhaskar (1975, 1998). There are, however, various schools of thought that could be termed critical realism and here we are providing an analysis of causality at a more fundamental level, relying on work that focuses specifically on this problem area.

A Framework for Causal Analysis in Design Science Research

Our analysis of causal mechanisms has pointed to six important perspectives for analysing causality that can potentially be used by researchers in design science research. Some of these perspectives are used commonly (if implicitly) by researchers and some are less common. But few design science researchers explicitly analyse and identify the causal claims upon which proposed design principles or theories are founded. In this section, we advance a framework that indicates how the different ways of thinking can be used to good effect.

First, we need to say something about the different types of artefacts that are dealt with, as the artefact type also influences the type of causal reasoning that is appropriate. Little discussion in the DSR literature distinguishes among the different classes of information systems produced. Recent research on design of organisations suggests a design distinction based on teleological goals. The same reasoning applies to the design of information systems intended to address different problem domains or different purposes. For example, the work of Germonprez et al. (2007) theorises about a class of artefacts that is intended to be modified in the context of use. This raises an interesting question regarding design theories for artefacts that are intended to be 'rigid' or explicitly not modifiable by the end user. In the first case, aspects of the system are emergent and therefore the *a-priori* causal analysis might be limited. In the second instance, the type of causal reasoning required will include causal analysis of how to prevent an outcome or how the absence of a feature might be a cause of something not occurring. A starting point for developing a framework for causal reasoning in DSR is a functional typology (Figure 3.1) of information systems such as that proposed by Iivari (2007), from which we can abstract dimensions for a causal framework.

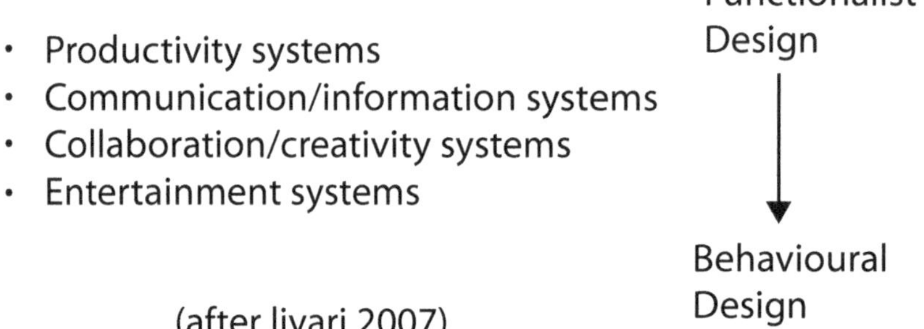

Figure 3.1 Teleological Abstraction of Information System Typology

This highly abstracted typology identifies a dimension along which all information systems fall. On one end are highly functionalist systems (Hirschheim and Klein, 1989) designed predominantly as productivity systems intended to achieve well-defined outputs with maximum efficiency from well-understood processes. As the processes, inputs, outputs and interactions are well known and understood, the causal connections and boundaries in the problem space are also well understood and the outcomes are highly predictable. Thus, specific types of causal reasoning are required. Characteristic of highly functionalist systems are tight coupling and strong component integration such as are found in accounting information systems and enterprise resource planning systems.

These systems can be contrasted with systems incorporating more behaviourally oriented design such as highly interactive collaboration systems or interactive entertainment systems, which privilege flexibility, creativity, adaptation to new problem domains and secondary design (Germonprez et al., 2007). This class represents design domains in which the users' behaviour and intentions are not only present, but are also required by the artefact-in-use. The contexts, tasks and users are diverse and variable and the systems are likely to evolve new patterns of in-situ use as they are modified. To obtain desired outcomes of system use requires types of causal reasoning that are enabling or probabilistic. Examples include design principles for learning systems or emergent knowledge processes (Markus et al., 2002). This distinction between design goals suggests a dimension of planned-emergent design, which forms one axis of our framework.

The other axis of our framework is formed by a distinction between the theorising that is done in designing artefacts (the interior prescriptive mode where artefacts are constructed to alleviate problems in the problem space) and the closely linked exterior descriptive mode, composed of the interactions of the artefact with its embedded context and its evaluation (Gregor, 2009). Although it is possible to conflate the interior and exterior modes of Gregor (2009) with the build and evaluate phases of Hevner et al. (2004), the distinction is important.

The interior mode focuses on theorising how artefacts can be designed and brought into being and is closely related to the build phase of Hevner et al. (2004). Here is where kernel theories are synthesised and causal connections are specified. Specifically, the abductive logic by which the explanation contained in the kernel theory (which is at a specific level of analysis and specific degree of generalisability) can explicitly define the connection between the principle and the expected outcome in the new design theory. The interior mode will often be iterative, with ongoing testing and experimentation helping to guide the design.

In contrast, the exterior mode focuses on the artefact-in-use after design is relatively complete or stable and the artefact is studied as part of a wider system, often by people other than the original designers. The exterior mode potentially includes all types of investigation, including measures of process or system output changes, user and management perception studies, phenomenological or hermeneutic studies of attached meaning and power structures or resistance. In this way, the exterior mode is differentiated from the evaluate phase of Hevner et al. (2004), which is predominantly focused on changes in efficiency, quality and efficacy. Knowledge gained from exterior mode research should include identifying causal connections for any research phenomenon related to the artefact-in-use. This might include negative outcomes, new problems or unexpected emergent behaviours that will inform the evaluation of the value of the artefact and, more importantly, inform future design activity.

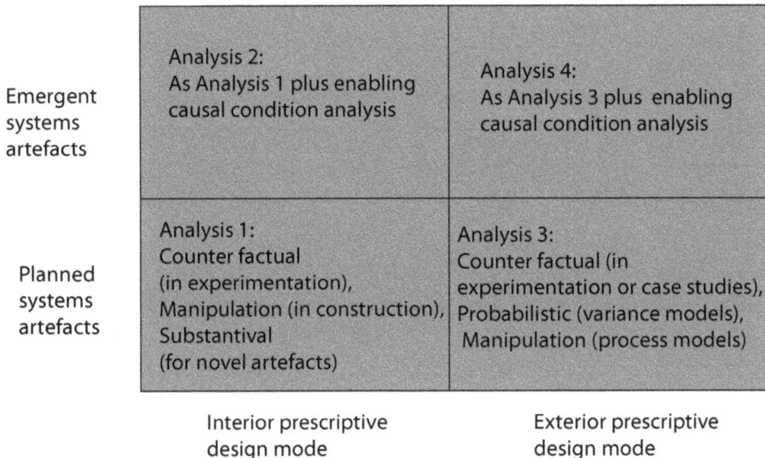

Figure 3.2 Types of Causal Analysis Useful in Design Science Research

Figure 3.2 shows the four cells that arise when these two dimensions are considered together, with indicative examples of appropriate causal reasoning given in each cell. The types of causal analysis suitable for each cell are now examined in more detail.

Analysis Cell 1: Interior design of planned systems

Examination shows that reasoning about causality in cells one and two differs in important ways from that in cells three and four, which are the cells associated with the traditional descriptive science approach. In these first two cells, the designer's thought processes in conceptualising a problem space and generating theoretical principles for potential solutions are themselves causal mechanisms. In the design of consequential management theory, Argyris (1996) suggests that the human mind functions as the designing system. This is what we term *substantival causality* (deliberative or mental causation). If we understood the direct causes or enabling conditions for human creativity and innovation, the design process could be manipulated to produce improved designs. But much design theory building is non-rational, abductive and unstructured. Reliance on kernel or reference theory to justify the 'idea' of the artefact is only part of the story; in many cases, we cannot say where the idea for the design came from, or why it is as it is, as human creativity and invention have come into play. Yet to evaluate the theoretical design principles and contribute to transferable knowledge, the design should be grounded in some type of reasoning that is amenable to causal analysis. To our knowledge, this type of causal analysis has not previously entered into discussion of design science research.

Many types of causal analysis can be used in both cell one and cell two, and DSR can be improved if they are explicitly applied. Manipulation analysis is used implicitly—that is, our team built this artefact and put it into use, with the implied prediction and expectation of a certain outcome. Here the analysis might consist simply of identifying what intervention will be created by the artefact and what system or behavioural change is expected as a direct result. This can be based on kernel theory, which demonstrates support for the causal linkage between manipulation and effect.

Counterfactual and probabilistic reasoning about causality are also used in an iterative design process. That is, the researcher constructs a prototype and experiments to see what results it causes, or does not cause, possibly in a probabilistic fashion. For instance, what percentage of test subjects prefer type A design to type B design? Iterative prototyping is inherently a process of refinement through identification of necessary and sufficient causal conditions. By adding or excluding specific physical conditions (affordances, conventions), psychological states (motivations, system explanations, user 'buy-in' through participatory design) and goal modification (final cause), the designer searches the design space for the constellation of causal conditions that increases the probability of production of the desired effects.

An example is given in Codd's work on the relational database model (Codd, 1970, 1982). Codd made claims about how fewer mistakes would occur with use of relational databases because users would not have to expend so much effort dealing with the complexity of repeating groups. This is counterfactual analysis: the removal of the artefact feature of repeating groups from the human-use process is the cause of fewer errors.

Analysis Cell 2: Interior design of emergent systems

Although it seems counterintuitive to conjoin design and emergence there is a strong impetus to create some types of artefacts whose functions, applications and behaviours are flexible, agile and emergent. In addition to the types of analysis supporting cell one artefacts, there is the need to consider enabling casual condition analysis. In this type of analysis, specific design principles are selected because of evidence that they will increase the probability that a desired outcome will be encouraged or supported. As specific emergent phenomena cannot be predicted, the principles that will improve the likelihood that general desirable characteristics (for example, flexibility, mutability, ability to be reconfigured) will emerge are selected. These might be *conditional causes* where the designer considers enabling (or disabling) environmental conditions that increase the probability of an outcome (Sloman, 2005). Examples include identification of causes that are likely to create perceived affordances, secondary design or combinatorial application of functions (for example, services). Principles such as component architectures, recognisable conventions and metaphors (Germonprez et al., 2007) suggest necessary but not sufficient causal conditions for the potential of emergent system behaviour. Counterfactual analysis can be applied in reverse to identify factors or processes that rigidly couple system components to the world, resulting in brittle, inflexible system use (Winograd and Flores, 1986).

The design work by Braa et al. (2007) is an example of theorising in this cell. They call their work action research but they offer design principles. For example, to create a new health standard in a context that is characterised as a complex adaptive system, one should actively create an attractor—one of a limited range of possible states about which the system will stabilise. Another example is in service-oriented systems, in which the user creates relationships among services by determining types and relevancy of data and outputs, and what things go together (Hovorka and Germonprez, 2008).

Analysis Cell 3: Observation of planned systems in exterior mode

The reasoning about causality in this cell can employ the methods of counterfactual analysis advanced by authors such as Shadish et al. (2002) for experimental and quasi-experimental work. For example, claims for the advantages of the relational database model in terms of the hypothesised reduction in programmer error and greater ease of use could be tested in experiments. Case studies can also use counterfactual analysis in pattern analysis. We turn again to Braa et al. (2007) who examined cases of attempts to develop health standards in several different countries. They analysed chains of events (process models) in each case but they also contrasted what happened and did not happen in each country (a form of counterfactual analysis).

Probabilistic analysis can be done using statistics, in what is often referred to as testing of variance models, accompanied by reasoning about why causal effects should hold and how other explanations for effects can be ruled out. In many cases, however, the reasoning from statistical analysis relies on correlations and analysis of covariance. Researchers should be more aware of statistical techniques recommended for attribution of causality (see Pearl, 2000). Further, claims for causality can be examined in terms of manipulation analysis when process models are examined.

Analysis Cell 4: Observation of emergent systems in exterior mode

Attribution of causality in this situation is difficult precisely because the outcomes were not actually designed for, but rather emerged from the in-situ use of the artefact. Yet as Gregor and Jones (2007, p. 326) note, 'the ways in which [artefacts] emerge and evolve over time and how they become interdependent with socio-economic contexts and practices' are key unresolved issues for design. Numerous researchers have noted that artefacts are often used in ways that were not intended due to tinkering or secondary design of the system (Ciborra, 2002; Hovorka and Germonprez, 2010; Romme, 2003) and the inability of designers to share the same model of the design space as held by the users (Dourish, 2001). As noted in cell two, here, design principles to enable or constrain emergent system behaviours can be designed into the artefact, but particular emergent characteristics cannot be predicted.

In the evaluation of emergent system behaviours, probabilistic counterfactual analysis might be possible and even desirable. Determination of what causal mechanism was present that enabled emergent behaviours broadens the scope and fruitfulness of design theory. In other instances of emergent behaviours, the design knowledge contribution might be in identifying mechanisms by which to extinguish or prevent behaviours. For example, secondary design of interfaces is not desirable in enterprise accounting systems or systems that require many information hand-offs. The principles for designing 'rigid' artefacts that are not amenable to secondary design are a largely unexplored area.

In concluding this section, we note that, not unexpectedly, in no cell was the first type of causal reasoning distinguished by Kim (1999) found to be relevant for socio-technical information systems. Because of the socio-technical complexity of designed and implemented information systems, we could find no example of causal reasoning that employed the logic of uniform and constant covering laws.[2] This observation has significant implications for the use of kernel theories empirically grounded in statistical evidence. As the kernel theories are only predictive in a probabilistic sense, derived design principles are frequently probabilistic. For DSR, this increases the knowledge creation burden on the evaluation phase, notably as counterfactual analysis can be used to identify the contexts or interactions in which the desired outcomes were not obtained.

Discussion and Conclusions

This essay has examined how causal reasoning can be employed in design science theorising. It has developed a framework with six types of potential causal analysis. The first four types are for event causation and include regularity analysis, counterfactual analysis, probabilistic analysis and manipulation analysis. A further two types are for agent causation and consist of substantival causation and enabling causal condition analysis.

Further, the essay develops a second framework that identifies the types of causal analysis that are suitable in different forms of DSR theorising. The orthogonal axes of this framework note distinctions on two different dimensions: 1) a planned versus emergent type of designed system; and 2) whether the work is in the interior prescriptive mode or the exterior descriptive mode of research. The four cells are labelled: 1) design of relatively stable planned systems; 2) design of emergent systems; 3) observation of planned systems in exterior mode; and 4) observation of emergent systems in exterior mode. The type of causal reasoning that can be used in each cell is described, with examples.

The question of substantival or mental causation in particular, although controversial, is worthy of attention because of its linkage to the truly novel

2 That the technical aspects of socio-technical systems are expected to behave in a uniform and predictable manner (for example, electronic circuitry) leads some researchers to reason in terms of covering laws. Cell one is where such reasoning, which we argue is very specific and limited, would appear.

artefacts that are a primary goal of design. Those reflecting on their research in DSR should consider how novel their artefact is. Genuinely novel and useful ideas and insights are likely to have greater impact. Codd's relational database work fell into this category. Reflection can distinguish novel innovations from new 'appliances' that might be more the result of normal industry practice, where knowledge of requirements plus knowledge of partial existing solutions that can be extended or adapted will cause an artefact to be produced in a fairly reactive fashion.

Our essay is significant because the topic of causal reasoning in DSR has received little, if any, attention. Our analysis has revealed ways of thinking about causality that have not been previously identified in the DSR literature. The position underlying the essay is that DSR can be better grounded by making clear the internal and theoretical warrants that underlie the theorising. Clarifying the causal claims invoked through kernel theories will improve theorising by providing criteria for kernel theory selection and delineating means of evaluation of the design theory based upon the assumed underlying causal claims. Even so, this essay also recognises that design theory can result from inspiration rather than theoretical or empirical grounding. But clear and explicit reasoning about causality and the different types of causal reasoning are a critical part of knowledge creation in the evaluation of design theorising. Causal reasoning has been shown to be an essential part of theory construction (Gregor, 2006). Our essay has practical implications because design theories underpin the construction of artefacts that are used in the real world, where the use of the artefacts can have consequences for both societal harm and societal good.

References

Abbott, A. (2001). *Time Matters: On theory and method.* Chicago, Ill.: University of Chicago Press.

Argyris, C. (1996). Actionable knowledge: design causality in the service of consequential theory. *The Journal of Applied Behavioral Science, 32*(4), 390–406.

Bhaskar, R. (1975). *A Realist Theory of Science*. London, UK: Verso.

Bhaskar, R. (1998). *The Possibility of Naturalism: A philosophical critique of the contemporary human sciences*. London, UK: Routledge.

Boland, R. J. (2002). Design in the punctuation of management action. In R. Boland (ed.), *Managing as Designing: Creating a vocabulary for management education and research*. Frontiers of Management Workshop, Weatherhead School of Management, 14–15 June.

Braa, J., Hanseth, O., Heywood, A., Mohammed, W., & Shaw, V. (2007). Developing health information systems in developing countries: the flexible standards strategy. *MIS Quarterly, 31*(2), 381–402.

Bunge, M. (1967). *Scientific Research II: The search for truth*. Berlin, Germany: Springer Verlag.

Bunge, M. (2008). *Causality and Modern Science*. [Fourth edn]. London, UK: Transaction Publishers.

Ciborra, C. (2002). *The Labyrinths of Information*. Oxford, UK: Oxford University Press.

Codd, E. F. (1970). A relational model of data for large shared data banks. *Communications of the ACM, 13*(6), 377–87.

Codd, E. F. (1982). Relational database: a practical foundation for productivity. *Communications of the ACM, 25*(2), 109–17.

Collins, J., Hall, N., & Paul, L. A. (eds) (2004). *Causation and Counterfactuals*. Cambridge, Mass.: MIT Press.

Dourish, P. (2001). *Where the Action Is: The foundations of embodied interaction*. Cambridge, Mass.: MIT Press.

Dourish, P. (2006). Implication for design. *Proceedings of the SIGCHI Conference on Human Factors in Computing Systems (CHI)*, Montreal, Canada, 541–50.

Fay, B. (1996). General laws and explaining human behavior. In M. M. Martin & L. C. McIntyre (eds), *Readings in the Philosophy of Social Science* (pp. 91–110). Cambridge, Mass.: MIT Press.

Fay, B., & Moon, B. (1996). What would an adequate philosophy of social science look like? In M. M. Martin & L. C. McIntyre (eds), *Readings in the Philosophy of Social Science* (pp. 21–35). Cambridge, Mass.: MIT Press.

Germonprez, M., Hovorka, D., & Callopy, F. (2007). A theory of tailorable technology design. *Journal of the Association for Information Systems, 8*(6), 351–67.

Goldkuhl, G. (2004). Design theories in information systems—a need for multi-grounding. *Journal of Information Theory and Application, 6*(2), 59–72.

Gregor, S. (2006). The nature of theory in information systems. *MIS Quarterly, 30*(3), 611–42.

Gregor, S. (2009). Building theory in the sciences of the artificial. *Proceedings of the 4th International Conference on Design Science Research in Information Systems and Technology (DESRIST)*, Philadelphia, Pa.

Gregor, S., & Jones, D. (2007). The anatomy of a design theory. *Journal of the Association for Information Systems, 8*(5), 312–35.

Hevner, A. R., Ram, S., March, S. T., & Park, J. (2004). Design science in IS research. *MIS Quarterly, 28*(1), 75–106.

Hirschheim, R., & Klein, H. K. (1989). Four paradigms for information systems development. *Communication of the Association for Information Systems, 32*(10), 1199–216.

Hooker, J. N. (2004). Is design theory possible? *Journal of Information Technology Theory and Application, 6*(2), 73–82.

Hooker, R. (1996). *Aristotle: The four causes—Physics II.3*. Retrieved from <http://www.wsu.edu:8080/~dee/GREECE/4CAUSES.htm>

Hovorka, D., & Germonprez, M. (2008). Identification-interaction-innovation: a phenomenological basis for an information services view. *Information Systems Foundation: Answering the unanswered questions about design research*. Canberra, Australia.

Hovorka, D., & Germonprez, M. (2010). Reflecting, tinkering, and tailoring: implications for theories of information systems design. In H. Isomaki & S. Pekkola (eds), *Reframing Humans in Information Systems Development* (pp. 135–49). Berlin, Germany: Springer.

Hovorka, D. S., & Germonprez, M. (2009). Reflecting, tinkering, and bricolage: implications for theories of design. *Proceedings of the Fifteenth Americas Conference on Information Systems*, San Francisco, Calif.

Hume, D. [1748] (2004). An enquiry concerning human understanding. In L. P. Pojman (ed.), *Introduction to Philosophy: Classical and contemporary readings*. New York, NY: Oxford University Press.

Iivari, J. (2007). A paradigmatic analysis of information systems as a design science. *Scandinavian Journal of Information Systems, 19*(2), 39–64.

Kim, J. (1999). Causation. In R. Audi (ed.), *The Cambridge Dictionary of Philosophy* [Second edn] (pp. 125–7). Cambridge, UK: Cambridge University Press.

Lee, A., & Baskerville, R. L. (2003). Generalizing generalizability in information systems research. *Information Systems Research, 14*(3), 221–43.

Little, D. E. (1999). Philosophy of the social sciences. In R. Audi (ed.), *The Cambridge Dictionary of Philosophy* (pp. 704–6). Cambridge, UK: Cambridge University Press.

Markus, M. L., Majchrzak, A., & Gasser, L. (2002). A design theory for systems that support emergent knowledge processes. *MIS Quarterly, 26*(3), 179–212.

Mill, J. S. (1882). *System of Logic Ratiocinative and Inductive: Being a connected view of the principles of evidence and the method of scientific investigation.* New York, NY: Harper & Brothers.

Norman, D. (1988). *The Design of Everyday Things*. New York, NY: Basic Books.

Pearl, J. (2000). *Causality: Models, reasoning and inference*. Cambridge, UK: Cambridge University Press.

Ramiller, N. C. (2007). Virtualizing the virtual. In K. Crowston, S. Sieber & E. Wynn (eds), *Virtuality and Virtualization* (pp. 353–66). Boston, Mass.: Springer.

Romme, A. G. L. (2003). Making a difference: organization as design. *Organization Science, 14*(5), 558–73.

Salmon, W. (1998). *Causality and Explanation*. New York, NY: Oxford University Press.

Schopenhauer, A. (1974). *On the Fourfold Root of the Principle of Sufficient Reason*. La Salle, Ill.: Open Court Publishing Company.

Shadish, W., Cook, T., & Campbell, D. (2002). *Experimental and Quasi-Experimental Designs for Generalized Causal Inference*. Boston, Mass.: Houghton Mifflin.

Sloman, S. (2005). *Causal Models: How people think about the world and its alternatives*. Oxford, UK: Oxford University Press.

van Aken, J. (2004). Management research based on the paradigm of the design sciences: the quest for field-tested and grounded technological rules. *Journal of Management Studies, 41*(2), 219–46.

Venable, J. (2006). The role of theory and theorising in design science research. *Proceedings of the First International Conference on Design Science Research in Information Systems and Technology*, Claremont, Calif., 1–18.

Winograd, T., & Flores, F. (1986). *Understanding Computers and Cognition: A new foundation for design*. Norwood, NJ: Ablex Publishing Corporation.

Woodward, J. (2003). *Making Things Happen: A theory of causal explanation*. New York, NY: Oxford University Press.

Part Two: Theories and Theorising in Practice

4. Theorising about the Life Cycle of IT Use: An appropriation perspective

Justin Fidock
RMIT University

Jennie Carroll
RMIT University

Abstract

This chapter argues that theorising about the whole life cycle of information technology (IT) use is underdeveloped. Theories that explain one or more phases of the life cycle of IT use are discussed and critiqued and a candidate theory for understanding the whole life cycle identified: the model of technology appropriation (MTA). The MTA incorporates many of the strengths of other models but has a shortcoming with respect to explaining how and why changes in patterns of use occur over the life cycle. To address this weakness, theories of change are examined and incorporated with the MTA. It is argued that the resulting model provides a more complete description and explanation of the IT use life cycle.

Introduction

Understanding and predicting the use of information systems (IS) are two of the central concerns for IS researchers and practitioners (Benbasat and Zmud, 2003; DeLone and McLean, 1992; Karahanna et al., 1999; McLean et al., 2002; Trice and Treacy, 1988). A system that is under-utilised, misused or avoided altogether will not achieve the intentions of its designers or those who have procured the system. Given its centrality for both researchers and practitioners, it is important to identify the ways in which researchers choose to theorise about use. This is because the choice of theory influences what is included or excluded from consideration. If a theory were not developed in the context of examining the whole use life cycle then its ability to explain the whole would likely be constrained. The central premise examined in this chapter is that theorising about the whole life cycle of IT use is underdeveloped. An important corollary of this premise is that understanding the life cycle as a whole will make additional contributions to IS research and practice beyond those provided by theory focused on only a portion of the life cycle. Furthermore, an enhanced understanding of the life cycle will assist in identifying the limits of applicability of partial views of the life cycle.

In this chapter the life cycle of IT use is briefly described. Theories used to explain use, and the life cycle of use more broadly, are then discussed and critiqued and a candidate theory for understanding the whole life cycle identified: the model of technology appropriation (MTA). The MTA is seen to incorporate many of the strengths of the other models but is somewhat lacking with respect to explaining the how and why of changes in patterns of use over the life cycle. Additional theories or motors of change are therefore introduced—teleology, dialectic and evolution—which are incorporated along with the MTA as a way of addressing this weakness. The chapter concludes by arguing that the revised MTA offers a richer and more complete description and explanation of the use life cycle than has been available hitherto.

The Life Cycle of IT Use

The life cycle of IT use describes the phases through which use of an IT artefact transitions—from the period prior to use through to continued or discontinued use. How the life cycle is represented is influenced by the ways in which use is conceptualised and examined. When use is conceptualised as the extent of use, the life cycle entails pre-use, initial use and continued use (see Figure 4.1). The extent of use is commonly assessed via self-reporting measures of the frequency or amount of use. Use is framed as a thing that changes in value but not in identity or character. Alternatively, the life cycle can be understood in terms of the nature of use and includes adaptive use and stabilised use (see Figure 4.2). The nature of use is viewed as potentially taking qualitatively different forms such as adaptation, stabilisation and appropriation and is often identified using qualitative methods (Carroll, 2004). The diagram in Figure 4.3 captures how use is represented and explored from the two perspectives combined.

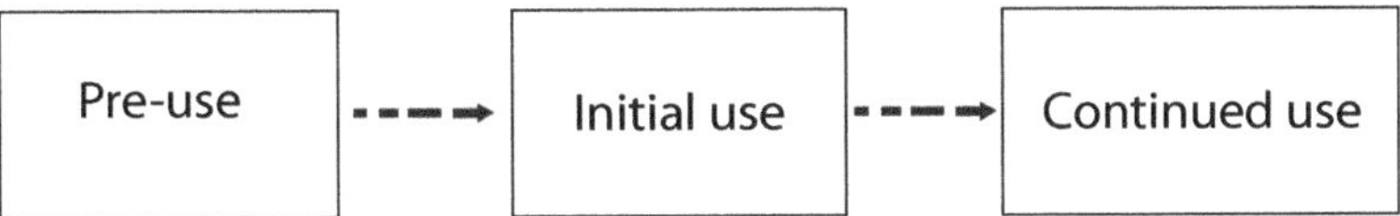

Figure 4.1 The Extent of Use Life Cycle

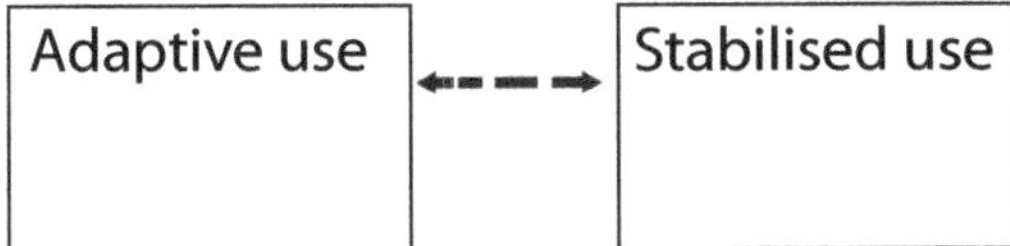

Figure 4.2 The Nature of Use Life Cycle

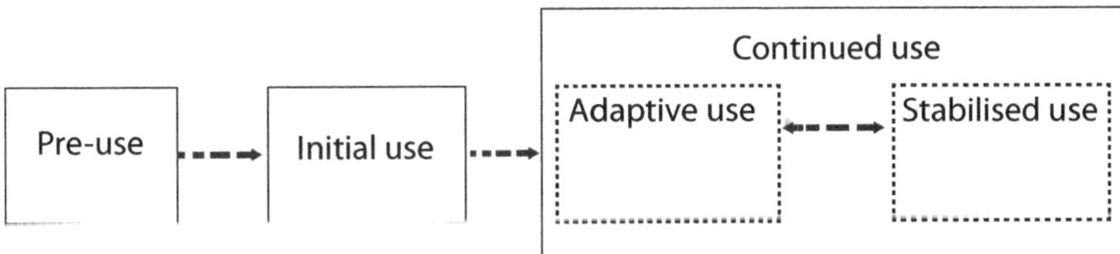

Figure 4.3 The Life Cycle of IT Use

Pre-use captures the period prior to a potential user starting to use a technology to support particular activities and practices (Bhattacherjee and Premkumar, 2004). This phase involves a user becoming acquainted with some of the features offered by the technology when first exposed to it, or following information about a proposed technology, as might occur via word of mouth. Attention is given to pre-use in situations where use is not well established, such as for new or prototype systems. In particular, researchers have attempted to predict future use by assessing users' intentions to engage in system use (behavioural intention) (Agarwal and Prasad, 1998; Davis et al., 1989; Karahanna et al., 1999; Mathieson, 1991; Taylor and Todd, 1995a, 1995b).

Pre-use is followed by initial use (Jasperson et al., 2005), during which time the user starts to employ features of the technology to support the conduct of tasks, such as might occur during a training course (Burton-Jones and Straub, 2006). In assessing initial use, researchers have used a variety of lean survey-based measures including frequency of use and time spent using the system (Adams et al., 1992; Al-Gahtani and King, 1999; Davis, 1989; Davis et al., 1989; Hubona and Geitz, 1997; Igbaria, 1990; Igbaria et al., 1989, 1997; Raymond, 1985; Roberts and Henderson, 2000). To a lesser extent, researchers have used hardware (DeLone and McLean, 1992) and software (Venkatesh et al., 2003) monitors to record actual system use.

Adaptive use occurs as a user engages in a more detailed exploration of the technology through applying the technology to support work practices within particular use contexts (Constantinides and Barrett, 2006; Desouza et al., 2007; Trigg and Bødker, 1994). Adaptations are made to the technology by a particular user to support their specific practices and use context, and adaptations are also made to user practices and the use context in response to the technology (Trigg and Bødker, 1994; Tyre and Orlikowski, 1994). Adaptations to the technology itself have been described using such terms as personalisation (Desouza et al., 2007), customisation (Desouza et al., 2007; Mackay, 1990a, 1990b; Trigg and Bødker, 1994), inventions (Desouza et al., 2007), reinvention (Johnson and Rice, 1984; Rice and Rogers, 1980; Rogers, 1995) and tailoring (Trigg and Bødker, 1994). Adaptations to work practices include work-arounds and improvisation (Hayes, 1999). Researchers also describe mutual changes in the technology and associated practices through such concepts as mutual adaptation (Boersma and Kingma, 2005; Carroll, 2004; Leonard-Barton, 1988; Majchrzak, et al. 2000), mutual adjustment (Rose and Jones, 2005), co-adaptation (Mackay, 1990a, 1990b), coevolution (Kim and Kaplan, 2006), structuring (Barley, 1986; DeSanctis and Poole, 1994) and the process of appropriation (Carroll 2004; Mendoza et al., 2008).

Stabilised use entails the routinisation of patterns of use (Rogers, 1995; Trigg and Bødker, 1994; Tyre and Orlikowski, 1994). Routines that develop might change in response to discrepant events or new discoveries by users (represented by the two-way arrow between adaptation and stabilisation) (Mendoza et al., 2005; Tyre and Orlikowski, 1994). Such stabilised use is also referred to as structured (Trigg and Bødker, 1994), integrated, appropriated (Carroll, 2004), infused (Sundaram et al., 2007) or institutionalised use (Orlikowski, 1992).

Continued use is an alternative to adaptive and stabilised use when the nature of use is not explored. It is the period of use that follows initial or first use (Hsieh et al., 2008; Karahanna et al., 1999; Pollard, 2003; Rogers, 1995; Shih, 2008; Venkatesh et al., 2003). Like the initial use phase, frequency of use and time spent using the system are often assessed.

Theories to Explain the Life Cycle

There are a large number of candidate theories that could be drawn on to assist in describing, explaining or predicting one or more aspects of the IT use life cycle. The theories discussed below have been selected based on their alignment with one or more of the key phases associated with the life cycle, but also based on their prominence within the IS research community. The key theories considered in this chapter are technology acceptance, innovation diffusion, structuration and the model of technology appropriation. A selection of studies drawing on each of these theories is summarised in Table 4.1, along with the phases of the use life cycle considered.

Table 4.1 Categorisation of Some Theories and Studies of Technology Use by Life-Cycle Phase

	Phase of IT use life cycle				
	Pre-use	Initial use	Adaptive use	Stabilised use	*Continued use*
Technology acceptance	✓	✓			✓
Davis (1989)		✓			✓
Davis et al. (1989)		✓			✓
Venkatesh et al. (2003)		✓			✓
Taylor & Todd (1995a, 1995b)	✓				✓
Innovation diffusion	✓				✓
Agarwal & Prasad (1998)	✓				
Karahanna et al. (1999)	✓				✓
Moore & Benbasat (1991)					✓
Structuration			✓	✓	
DeSanctis et al. (2000)			✓	✓	
Model of technology appropriation	✓	✓	✓	✓	
Carroll et al. (2003)	✓	✓	✓		
Mendoza et al. (2005)	✓	✓	✓	✓	
Mendoza et al. (2008)	✓	✓	✓	✓	

Table 4.1 suggests that very few researchers have considered use over the whole life cycle (four phases), or even over three phases. The majority of research on the use of IT artefacts only provides a partial view of use across the life cycle. Technology acceptance and innovation diffusion perspectives attend to the extent of the use life cycle (see Figure 4.1). Structuration focuses on the nature of the use life cycle (see Figure 4.2). Only studies drawing on the MTA cover the whole life cycle, from pre-use through to stabilised use. Our attention now turns to the descriptive and explanatory potential offered by each of these theoretical perspectives.

Technology Acceptance

The theoretical grounding for much of the research into user acceptance comes from the technology acceptance model (TAM) developed by Davis et al. (1989), Venkatesh et al. (2003) and Venkatesh and Bala (2008). This model is an adaptation of the theory of reasoned action (TRA), which sees beliefs and attitudes as antecedents of future behavioural responses, such as actual system use (Ajzen, 1985; Davis, 1993). TAM differs from the TRA by identifying the role of external variables more explicitly and by identifying two particular belief constructs as particularly relevant in the IS domain—namely, perceived usefulness and perceived ease of use. These beliefs are seen—either directly or indirectly and via attitudes towards using the technology—to shape users' intentions to employ a system, which in turn determines system use, which is usually conceptualised as the extent of use (Davis, 1989; Davis et al., 1989). Figure 4.4 presents TAM as represented by Davis et al. (1989). An important feature of the model is its focus on prospective users, as evidenced by the definition of perceived usefulness as 'a prospective user's subjective probability that using a specific application system will increase his or her job performance within an organisational context'. In subsequent studies the model and associated variables have been applied to both prospective and current users of systems, through changing the tense of scale items (for example, Venkatesh et al., 2003).

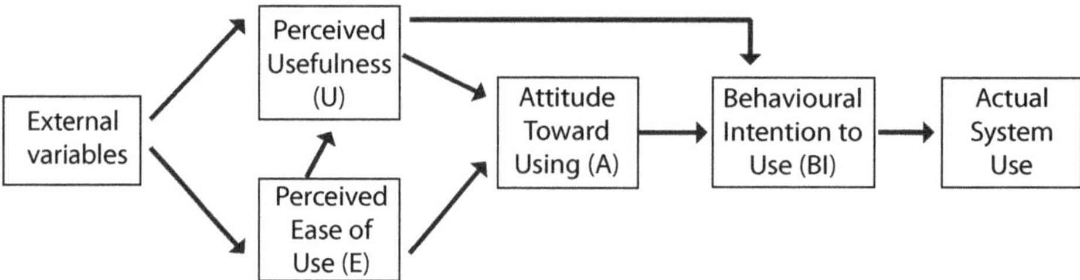

Figure 4.4 The Technology Acceptance Model

Source: Davis et al. (1989, p. 985).

The research by Davis (1989) and colleagues (Davis et al., 1989) on technology acceptance and its antecedents (perceived usefulness and ease of use) has had, and continues to have, an enormous influence on IS research. This is a positive in the sense that it has laid the foundation for a cumulative tradition whereby there has been widespread and persistent use of the two key variables—perceived ease of use and usefulness—in a diverse range of use contexts and technologies. In addition, statistical methods such as regression analysis and structural equation modelling that have been employed in these numerous studies to test hypothesised links between these variables and use afford researchers statistical control that can be construed as a form of experimentation

(Lee, 1999). This can assist in cutting through the complexity of the phenomena of interest. Such statistical experimentation has assisted in confirming the central role of perceived usefulness, and to a lesser extent perceived ease of use, in predicting behavioural intention and use. The parsimony of TAM, however, and its apparent power in explaining a fair portion of the variance, appear to have led to an overemphasis on the extent of variance explained rather than the untidy and messy unexplained variance. A focus on pre-specified variables and the explained variance in models such as TAM leads to a neglect of context, which removes the capacity to understand the 'subtle nuances of interaction that are critical in apprehending what is really occurring' (Pfeffer, 1982, p. 75). As a result, whilst levels of perceived usefulness and ease of use would assist managers with understanding influences on intentions and use in a general sense, the information provided would say little about the specific contextual factors operating on users of the particular technology. Even when attention is constrained to explained variance there are methodological issues associated with questionnaires (Sharma and Yetton, 2001; Straub and Burton-Jones, 2007). A meta-analysis by Sharma and Yetton (2001, p. 1170) found that 'the average correlation between "perceived usefulness" and "use" is 0.26 in studies employing behavioral measures of use and 0.56 in studies employing perceptual measures of use'. This effect, whereby perceptual or self-reporting measures of use lead to higher correlations, is referred to as self-generated validity (Taylor and Todd, 1995b). It is proposed that 'when survey respondents are asked about issues to which they have given very little prior thought, they...are apt to use answers to earlier survey questions as the bases for responses to later questions, resulting in inflated causal linkages' (Taylor and Todd, 1995b, p. 171). So, rather than describing cognitions and behaviours, researchers run the risk of creating and even changing them (Ogden, 2003; Pfeffer, 1982; Taylor and Todd, 1995b).

The development of TAM was driven by a desire to predict and explain human behaviour (Davis et al., 1989). TAM and associated acceptance models such as UTAUT are cognitive-rational theories that assume system use is driven by the intentionality of users, with users' intentions being informed by their beliefs and attitudes towards the technology of interest (Pfeffer, 1982). To the extent that behaviour is driven by intention then such models have some utility; however, this class of theories has been widely criticised in the social and management sciences more broadly (Abraham and Sheeran, 2004; Ogden, 2003; Pfeffer, 1982). Such theories have been criticised for

- not helping to explain the variety of things people use, such as the range of features employed on a system, versus just explaining a particular behaviour of interest, such as extent of system use (Abraham and Sheeran, 2004)
- creating and shaping rather than describing users' cognitions and behaviours (Ogden, 2003; Pfeffer, 1982; Taylor and Todd, 1995)

- often assuming unidirectional causality and the associated implication that beliefs and attitudes come before behaviour, despite evidence that behaviour also shapes attitudes and beliefs (Mintzberg and Westley, 2001; Pfeffer, 1982); people do not always think or choose before taking action, such as when engaged in habitual behaviours; when asked about the reasons for their behaviours this could reflect retrospective rationality, not prospective rationality as is assumed in such models (Pfeffer, 1982)
- providing a weak source of explanation as to why the relationships identified are significant (Hovorka et al., 2008; Pfeffer, 1982). As Hovorka argues, 'a falling barometer allows inference that there has been a drop in air pressure, but the drop cannot be explained by referring to the barometer' (p. 32). In the context of TAM, the source of explanation is the set of variables statistically identified as linked to intentions and behaviours. Such statistical inferences only afford partial understanding of the phenomena of interest. Whilst the variables might predict use, their ability to explain use is limited.

Another limitation of user acceptance models like TAM and UTAUT is that time is viewed as part of the background (Van de Ven and Poole, 2005). The amount of time is uncritically applied as an indicator of experience, familiarity and routinisation (Venkatesh et al., 2003), or the interest in time is limited to providing distance between two measurements so as to determine the strength of the causal relationship between behavioural intention and system use (for example, Davis, 1989; Taylor and Todd, 1995b). These limitations raise serious questions about the utility of TAM and related models for understanding the life cycle of use, and it is for these reasons, and others, that there have been widespread calls to go beyond TAM (Bagozzi, 2007; Baron et al., 2006; Benbasat and Barki, 2007; Carayannis and Turner, 2006; Dishaw and Strong, 1999; Goodhue, 2007; Hirschheim, 2007; Lucas et al., 2007; Mendoza et al., 2005; Schwarz and Chin, 2007; Silva, 2007; Straub and Burton-Jones, 2007). Nevertheless, such models can assist in drawing inferences about the salience and strength of particular influences on intentions and use, which might be particularly relevant at the pre-use and initial use phases when users might be expected to be more driven by intentions (Venkatesh et al., 2003).

Diffusion of Innovation

The literature on the diffusion of innovations is diverse and populated by a variety of different models that address individual and organisational decision points and activities (Cooper and Zmud, 1990; Hage and Aiken, 1970; Johnson and Rice, 1984; King, 1990; Kwon and Zmud, 1987; Rice and Rogers, 1980; Rogers, 1995; Wolfe, 1994). One researcher in particular has, however, dominated research into the diffusion of innovations for decades: Everett Rogers. In his book *Diffusion of Innovations* (1995), Rogers presents models of the innovation diffusion process that emphasise either the individual or the organisational processes, as well as models of adoption and implementation. Of particular relevance here is his model that describes the innovation diffusion process for individuals (see Figure 4.5). The model has five stages

- knowledge: the stage where a potential adopter becomes aware of an innovation and develops some understanding of its capabilities
- persuasion: the stage where the formation of either positive or negative attitudes towards an innovation occurs
- decision: the stage where a person decides either to adopt or to reject an innovation
- implementation: the stage where a person puts an innovation to use
- confirmation: the stage where either the innovation decision is reinforced or an earlier decision to adopt or reject a system is reversed.

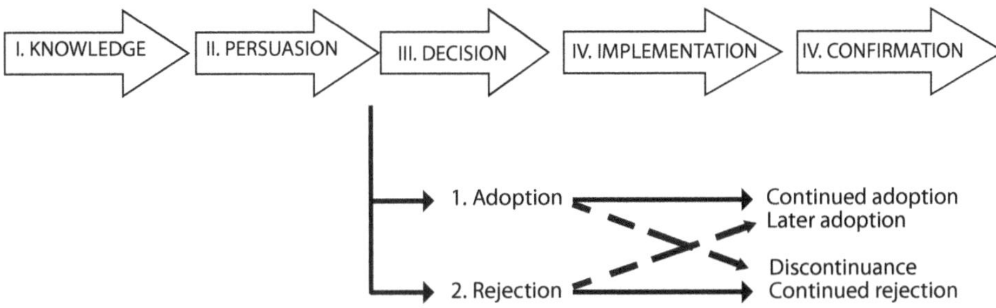

Figure 4.5 A Model of Stages in the Innovation Decision Process

Source: Rogers (1995, p. 163).

In addition to these stages, Rogers also identifies five key attributes of innovations that influence user adoption at the persuasion stage

- relative advantage: the extent to which an innovation is viewed as better than its predecessor
- compatibility: the extent to which an innovation is viewed as consistent with the extant needs, values, beliefs and experiences of potential adopters
- complexity: the extent to which an innovation is viewed as difficult to use
- observability: the extent to which the impacts of an innovation are visible to others
- trialability: the extent to which an innovation can be experimented with prior to the adoption decision.

The above model and associated five key attributes have been drawn on extensively in the IS literature (for example, Agarwal and Prasad, 1998;

Karahanna et al., 1999; Kautz and Larsen, 2000; Kraut et al., 1998; Lin and Lee, 2006; Moore and Benbasat, 1991; Pollard, 2003; Prescott and Conger, 1995; Rice and Rogers, 1980; Shih, 2008). They support exploration of influences prior to, during and after the adoption decision. Furthermore, the model draws a distinction between initial use, during which time the innovation is put to use (implementation stage), and continued/discontinued use (confirmation stage) (Pollard, 2003).

Rogers' model and other models of the innovation adoption and diffusion process (Carayannis and Turner, 2006; Cooper and Zmud, 1990; Hage and Aiken, 1970; Johnson and Rice, 1984; Kwon and Zmud, 1987; Nord and Tucker, 1987; Saga and Zmud, 1994) are life-cycle models. As such, change is explained by reference to the sequence of phases through which the system of interest passes (Van de Ven and Poole, 1995). Such life-cycle models tell us little, however, about the underpinning mechanisms that generate the behaviour observed, and, with some exceptions (for example, Rogers), tend to be analytically or meta-analytically derived rather than emerging from first-hand longitudinal research on the adoption and diffusion of technologies.

A counter to this argument might be to say that explanation is provided by assessing users' perceptions of a system using the five key attributes. While studies investigating the influence of these attributes help to identify important influences on adoption and use, they often do so from the perspective of cross-sectional or 'factor'-based research (for example, Agarwal and Prasad, 1998; Lin and Lee, 2006). As a consequence of this methodological constraint the clearly process-orientated nature of innovation diffusion theories can only be partially examined and explained (McMaster and Wastell, 2005). A further issue with these five key attributes is the assumption that they primarily operate at the persuasion stage, when there is clear evidence that such affective attributes operate after the adoption decision (Karahanna et al., 1999).

There are two additional criticisms that have been directed at the innovation diffusion literature. Historian David Edgerton argues that views of innovation tend to be centred on innovations as they are emerging, not after they have long been in use: 'Even as new technologies revolutionize everything from health care to media to warfare, it's important to remember that our world runs primarily on products and technologies long in use' (Baker, 2007). Rogers (1995, p. 172) also criticises innovation research for having a strong pro-innovation bias, going on to say that 'investigation of rejection behavior of all kinds has not received much scientific attention'.

Diffusion of innovation research provides a basis upon which to investigate the life cycle of use; however, it does so from a perspective that privileges the new over the taken for granted, adoption over rejection, and factor over process.

Structuration

Information systems researchers have extensively drawn on structuration theory—in particular, Anthony Giddens' structuration theory (1986)—to assist in explaining the interactions between technology and people embedded in social contexts, such as organisations (Jones and Karsten, 2008). Giddens was concerned with transcending the dichotomous logic associated with dominant traditions within social theory that privileged either the agency of individuals or the structures that limit human choices and action, such as properties of society. Structurational theories seek to reconcile tensions between individual and societal-level explanations of social phenomena through seeing both as being mutually constituted (Jones and Karsten, 2008). Social phenomena, from a structurational perspective, are the product of both structure and agency: 'human agents draw on social structures in their actions, and at the same time these actions serve to produce and reproduce social structure' (Jones and Karsten, 2008, p. 129). These structures, or more particularly the structural properties of social systems, consist of rules and resources used by individuals in their interactions. 'These rules and resources mediate human action, while at the same time they are reaffirmed through being used by human actors' (Orlikowski, 1992, p. 404). Whilst this process of mutual constitution enables change in social systems, Giddens' argued that continual change is countered by a desire amongst social actors for ontological security, a belief that their personal or professional identities will be maintained and that actions and events in the surrounding social and material environment will not undermine their identity. Predictability and routinisation are therefore of central importance in maintaining the identities of social actors (Jones and Karsten, 2008).

Giddens' structuration theory has been translated into IS in a variety of ways. The appeal of this theory for IS researchers is that it provides a means of adopting a non-dichotomous logic (Pozzebon, 2004). For IS researchers adopting a structurational perspective, the structure/agency dichotomy is overcome by framing the relationship between technology and humans as the process through which humans shape and are shaped by IT artefacts (Orlikowski, 1992; Poole and DeSanctis, 1990). One particularly dominant translation of Giddens' theory within IS is Poole and DeSanctis's adaptive structuration theory (Poole and DeSanctis, 1990).

Adaptive structuration theory (AST) emerged out of research focused on social interactions and processes associated with the use of group decision support systems (DeSanctis and Poole, 1994; Poole and DeSanctis, 1990). DeSanctis and Poole (1994) present a number of different propositions associated with their modification of structuration theory for IS; however, two propositions are particularly salient as they bring to the surface three key concepts associated with AST (highlighted in italics).

- Advanced information technologies (AITs) 'provide social structures that can be described in terms of their features and spirit. To the extent that AITs vary in their *spirit* and *structural feature* sets, different forms of social interaction are encouraged by the technology' (p. 128; italics added).
- 'New social structures emerge in group interaction as the rules and resources of an AIT are *appropriated* in a given context and then reproduced in group interaction over time' (p. 129; italics added).

Structural features represent particular capabilities, or rules and resources, provided by the system. Structural features 'govern exactly how information can be gathered, manipulated and otherwise managed by users' (p. 126). Underlying these structural features is the way in which users should act when employing the system, referred to as the spirit. The spirit of a technology reflects, amongst other things, the designers' intentions; however, the ways in which users appropriate or implement the technology are not necessarily determined by the technology design. The particular structural features selected by users represent only a subset of those embedded within a technology. Users therefore are able to appropriate the capabilities of a system in a wide variety of ways (DeSanctis and Poole, 1994). Appropriation is here understood to be the 'immediate, visible actions that evidence deeper structuration processes' (p. 128). Analysis of appropriation moves therefore provides a way of examining the underlying social processes.

One of the concerns about AST is its view of technology as encapsulating social structures in the form of structural features and spirit. This view runs counter to the position adopted by Giddens, who argued that social structures do not have an existence independent of the action of humans (Markus and Silver, 2008). A further concern is the apparent anthropomorphism of the spirit concept, which is described as somehow conveying or embodying the designers' intentions (Markus and Silver, 2008). Another issue with AST is that research that has operationalised one or more of its key constructs has failed to identify strong relationships with common influences such as ease of use and usefulness; instead the findings are inconsistent and only moderate in strength (Chin et al., 1997; DeSanctis and Poole, 1994; DeSanctis et al., 2000; Gopal et al., 1992; Salisbury et al., 2002). AST also appears to have been predominantly applied to group or collaborative information systems such as group decision support systems and computer-mediated communication (Jones and Karsten, 2008), rather than less socially mediated systems.

A more general criticism of structurational approaches is that readers experience difficulties in readily understanding the meaning of the text. Information systems is an applied discipline and it has been argued that the accessibility of theories is an important consideration in judging relevance (Rosemann and Vessey, 2008). Structuration theory and AST are frequently difficult to comprehend and employ concepts in ways that bear little relation to their more common forms of use. For example, Giddens' definition of structure, defined in terms of rules and resources, is particularly idiosyncratic (Jones and Karsten, 2008).

Researches drawing on Giddens' theory, in particular his views on the importance of ontological security, follow him in privileging routinisation and the maintenance of individual identities over adaptation and transformation (Chu and Robey, 2008). While adaptation has its place, it is subordinate to stability. This position is consistent with empirical research suggesting that over time adaptations become structured (for example, Carroll, 2004; Trigg and Bødker, 1994; Tyre and Orlikowski, 1994). Research has, however, also found that adaptations can again occur (Mendoza et al., 2005, 2007; Tyre and Orlikowski, 1994).

Model of Technology Appropriation (MTA)

The model of technology appropriation (MTA) was developed by Carroll et al. (2002c) to assist with understanding the process of appropriation through which technology is evaluated by people over time and adopted, adapted and incorporated into their work practices, and through which the design of technology is completed through use (Figure 4.6).

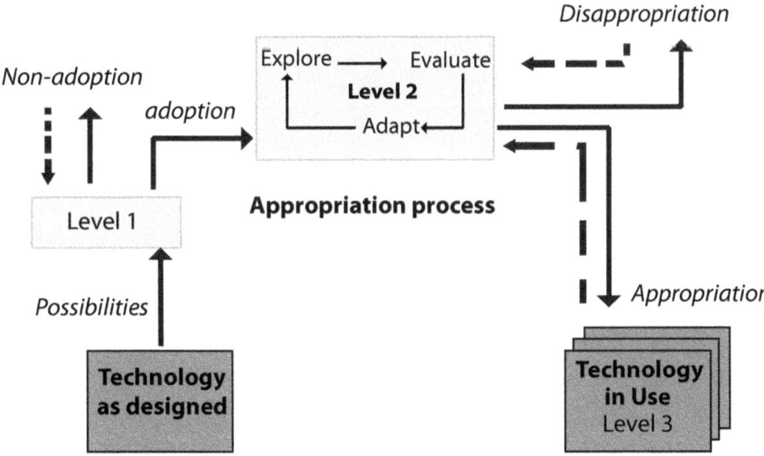

Figure 4.6 The Model of Technology Appropriation

Source: Carroll (2004, p. 5).

The MTA is intended to be a generic model of technology appropriation that can be tailored for particular technologies and user cohorts (Carroll, 2004). It has been employed to assist in describing the appropriation of mobile phones, bibliographic software, Short Messaging Service (SMS), email, customer relationship management software, open source software and a learning management system (Carroll et al., 2002a, 2003a, 2003b, 2005; Herszfeld et al.,

2003; Heung, 2002; Mendoza et al., 2007; Nor Zairah and Rose Alinda, 2007). Throughout the process of appropriation various influences shape the beliefs, attitudes and behaviours of users towards the technology.

In the model there are three levels of evaluation that correspond with different phases of the appropriation process (Carroll et al., 2002a). As can be seen in Figure 4.6, when first encountering a technology during the initial exposure phase, users are confronted with the technology as intended by its designer, or 'technology as designed', which 'has features, capabilities and an underlying theory or spirit' about how the technology should be employed (p. 3). From the users' perspective, the technology presents a variety of possibilities for addressing their particular concerns, which might or might not align with those identified by the designers. During users' initial exposure to the technology a series of influences shapes their evaluations and decisions whether or not to adopt the technology. In the case of an information system, influences on users might include the graphical user interface and system functionality. The outcome of this level-one evaluation is the establishment of certain expectations about what the technology can deliver, which leads to either non-adoption or the user choosing to persist with exploring the technology thereby continuing the appropriation process. In the case where the user chooses not to adopt the technology there might be circumstances that cause them to re-evaluate the technology at some later time (represented by the dashed arrow from non-adoption to level one in Figure 4.6).

At the next phase of the appropriation process users evaluate the technology more deeply through exploring and using the technology (level-two evaluation) (Carroll, 2004; Herszfeld et al., 2003). They come to learn how the technology can support their practices through the provision of particular functionality. As users explore and learn about the technology they also adapt their practices associated with the technology as well as adapting the technology itself. During this adaptation phase there is again a variety of influences that serves to encourage or discourage continued appropriation—for example, the extent to which the technology enhances the user's performance.

In the final phase a state of appropriation or stabilisation is reached, whereby the practices around the use of the technology become routine, and no further adaptations to the technology occur (Carroll, 2004). The technology becomes integrated with work practices, is part of users' taken-for-granted experience of work, and is just another part of the work landscape, referred to as 'technology in use'. It is during this phase that the design can be said to be fixed, although this might not be permanent (Mendoza et al., 2005) (see dotted arrow back to level two). The state of appropriation is maintained as long as users' ongoing evaluation of the 'technology in use' continues to reinforce persistent use. The particular patterns of use that users settle on are assumed to vary across individuals, conveyed by the tiled boxes associated with 'technology in use' in Figure 4.6. These level-three evaluations are shaped by various influences—for example, the attitudes and behaviours of one's peer group towards the technology or the performance of the technology. Users' persistent use and ongoing incorporation of the technology with their work practices are, however, subject to modification if their evaluation of the technology changes. If this occurs then users might return to level two and the technology could be disappropriated or rejected.

The two primary strengths of the MTA are that it covers the life cycle from pre-use through to stabilised use, and it is a readily accessible and quite parsimonious model. It also explicitly distinguishes between adaptation and stabilisation (although the latter is referred to, somewhat confusingly, as appropriation), as well as incorporating the adoption decision and the possibility of technology rejection occurring after a period of use. In addition, the model highlights the role of influences in shaping users' evaluations and associated patterns of appropriation throughout the process, although, due to the generic nature of the model, these influences are not specified and are assumed to vary across technologies and cohorts (Carroll, 2004). Nevertheless, in describing the model, Carroll (2004) argues that the features of the technology are particularly important when first encountering a technology, with system usefulness becoming more salient as users apply the technology in context.

The model shares concepts from the models and theories described earlier. The concept of 'spirit', also used by DeSanctis and Poole (AST), is drawn on to help describe the 'technology as designed'. The MTA incorporates the concept of mutual shaping or adaptation, like AST. The MTA includes the decision to adopt as an outcome of initial exposure, as does Rogers (1995). The MTA therefore provides a means of describing use of technology over time in a way that is quite nuanced, as well as being consistent with prior research; however, the model emphasises description over explanation and prediction, which is not unexpected given that the model emerged from empirical research. It therefore does not provide much in the way of insights as to the underpinning or generative mechanisms that shape users evaluations and patterns of appropriation, other than to flag the need to identify the particular influences operating on users' evaluations of a particular technology.

Identification and Development of a Theory for Understanding the Use Life Cycle

Each of the theories or models considered above has a variety of strengths and weaknesses that can assist in understanding one or more aspects of the use life cycle, which are summarised in Table 4.2.

Table 4.2 Relative Strengths of Theories for Understanding the Use Life Cycle

Relative strength	Technology acceptance	Innovation diffusion	Structuration	MTA
Consideration of whole use life cycle	Low	Medium	Low	High
Accessibility/parsimony	High	High	Low	High
Understanding of dynamics of influences on and patterns of use	Low–medium	Low–medium	Medium	High
IT artefacts considered at micro-level (not black boxed)	Low	Low	High	High
Ability to explain process of change (explanatory power)	Low–medium	Low–medium	Medium–high	Low–medium
Inductive/empirical basis for model	Low	Medium	Low	High
Applied to range of user cohorts and use contexts	High	High	High	Medium
Consideration of mutual adaptation/bidirectional causality	Low	Medium	High	High
Focus on adaptation and stabilisation	Low	Medium	High	High
Stabilisation not privileged over adaptation	Low	Low	Low	Medium
Consideration of context	Low	Low	High	High
Consideration of minimal use, rejection behaviour	Low	Medium	Medium	High
Cumulative tradition	High	High	Medium	Low
Understanding of heterogeneity of use across individuals	Low	Low–medium	Medium	High

Understanding the whole life cycle of IT use is of central concern in this research. Only the MTA provides coverage across the entire life cycle from pre-use through to stabilisation. The model also facilitates understanding of the dynamics of influences on and patterns of use, and does so in a way that is readily accessible and quite parsimonious. Furthermore, it considers technology and relationships with users at the micro-level of analysis. It is for these reasons, as well as the additional strengths listed in Table 4.2, that the MTA is the most promising candidate theory for understanding the whole life cycle of IT use. There are, however, two areas where the MTA is less strong: its ability to explain the how and why of the appropriation process, and the limited cumulative tradition. The latter issue is addressed through drawing on the MTA in this chapter, thereby contributing to the ongoing establishment of a cumulative tradition. The former concern about explanatory power is addressed quite well in structurational models like AST, but such power is diminished by their low accessibility and parsimony. What are required are theories of change that complement the MTA by enhancing the capacity to explain the appropriation process in a way that is readily accessible.

Van de Ven and Poole (1995) reviewed theories of change in the biological, physical and social sciences. They identified four 'ideal type' theories of change: life cycle, teleology, dialectic and evolution. The MTA provides largely a life-cycle perspective on the change process whereby change is explained in terms of a sequence of phases through which the system of interest passes. The progression through the phases is presumed to follow a certain immanent logic or sequence that is pre-programmed. While the environment influences how the entity expresses itself, such as the particular patterns of adaptation and stabilisation, as well as their timing, these types of change events are nevertheless mediated by the immanent logic, or what Van de Ven and Poole (1995) referred to more broadly as the generative mechanism. The use of a life-cycle perspective provides a way of generating rich descriptions of the entity of interest. It is, however, somewhat limited with respect to explaining how and why the entity of interest changes or remains stable over time. This constraint can be overcome by juxtaposing additional theories of change and their associated generative mechanisms.

A teleological perspective frames change as being driven by the purposeful pursuit of goals (Van de Ven and Poole, 1995). The generative mechanism is the enactment of goals, which in the IT domain would be undertaken by users or organisations. Users or organisations are seen to act as intentional agents working to achieve the fulfilment of their goals. Furthermore, these agents are presumed to be adaptive and creative in formulating and enacting their goals. Unlike life-cycle theories there is no prescribed sequence. Instead, there is 'a repetitive sequence of goal formulation, implementation, evaluation, and modification of goals based on what was learned or intended by the entity' (p. 516). Theories in IS that draw on cognitive rational theories, such as many theories of acceptance and innovation diffusion, similarly assume that change is driven by the intentionality of users, with users' intentions being informed by their beliefs and attitudes towards the technology of interest (Pfeffer, 1982).

Dialectic theories explain stability and change by reference to the tension that exists between opposing or contradictory forces, such as that between advocates of the status quo, the thesis and those promoting change—the antithesis (Van de Ven and Poole, 1995). The types of outcomes resulting from tensions can be understood in terms of maintenance, substitution or synthesis. Maintenance describes the continuance of the status quo, with the thesis dominating the antithesis. Substitution occurs when the thesis is replaced with the antithesis. The third possible outcome is a synthesis between the thesis and antithesis—an outcome that is distinct from its constituent elements. The generative mechanism or motor of change in dialectic theories is the tension or conflict that exists between opposing forces. None of the theories considered here clearly represents or draws on a dialectic perspective, although there are examples of such theories being employed in IS (Cho et al., 2007; Myers, 1994; Robey and Boudreau, 1999; Robey et al., 2002). Giddens' structuration theory incorporates dialectic elements by identifying the possible tensions that exist between human agency and the structural properties of the contexts within which humans are embedded. The synthesis from this tension is the process of mutual constitution of agency and structural properties; however, it is not clear how agency or structure could exist independently of the other, as is the case in the dialectic theories described by Van de Ven and Poole (1995).

Evolutionary theory views and explains change as occurring through a continuous process of variation, selection and retention (Van de Ven and Poole, 1995). Variation comes about due to random or unpredictable changes or events. Selection occurs through competition for scarce resources in the environment. Retention refers to maintenance of an entity's form; it serves to counteract the 'self-reinforcing loop between variations and selection' (p. 518). An evolutionary perspective therefore captures the tension between change and inertia associated with the status quo or temporary stabilisations. The role of unpredictable discrepant events in explaining additional adaptations in the research by Tyre and Orlikowski (1994) is an example of research that is consistent with an evolutionary perspective. There are studies that have also more explicitly drawn on one or more aspects of evolutionary theory in the IS domain such as coevolution (Fidock, 2002; Kim and Kaplan, 2006) and punctuated equilibrium (Lyytinen and Newman, 2008; Mendoza et al., 2007; Sabherwal et al., 2001).

The teleological, dialectic and evolutionary lenses, together with the life-cycle perspective offered by the MTA, offer the potential for providing greater understanding of the life cycle of IT use than would be provided by drawing on only one theoretical perspective. This is because particular theoretical perspectives, as metaphorical devices or lenses, draw attention to particular features or qualities whilst also leaving out others. It is the integration and juxtaposition of these theories to develop new theory that has stronger and broader explanatory power than the initial perspectives (Van de Ven and Poole, 1995). Through incorporating the theories of change, the explanatory power of the MTA is enhanced. This is summarised in Figure 4.7. Each of the theories of change offers alternative and complementary explanations of why appropriations change over time. The contribution of each theory of change to enhancing the MTA is now examined through the case of electronic mail in the Defence Science and Technology Organisation (DSTO). Due to space limitations, consideration is limited to the separate contributions of each change theory. Future work will examine the ways in which the theories can be combined and sequenced to further enhance explanation.

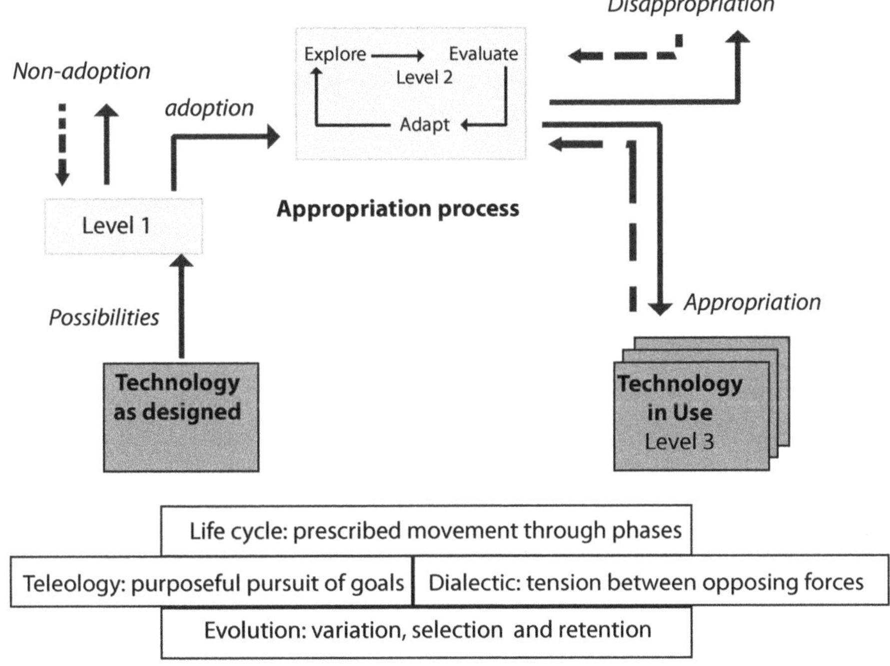

Figure 4.7 An Enhanced Theory for Explaining the IT Life Cycle

Source: Adapted from Carroll (2004, p. 5).

Explaining the IT Use Life Cycle: The case of email

Life Cycle

In DSTO, email is a mature and pervasive technology that is thoroughly incorporated with peoples' practices to become a 'technology in use'. Furthermore, the amount of exposure of individuals to this technology was substantial and ranged from four to 28 years. In the MTA, 'technology in use' is associated with stabilisation in patterns of appropriation. There was evidence of stability, such as frequency of filing messages and the way email was managed. Also consistent with the MTA, there was evidence of adaptations occurring to the default configuration of *MS Outlook 2003* after its introduction, such as turning off the reading pane or turning off the message reminders, followed by stability in these reconfigurations. There were also instances of people making further changes to their configurations as a result of reflecting on how their current configuration might not readily support desired practices; deciding, for example, to turn off the message reminders after discussing this feature with the researcher. The MTA therefore provides a way of describing phases through which users might pass: adaptation, followed by stabilisation then renewed adaptation.

Another feature of the MTA is the proposition that users' trajectories with respect to how they come to adopt, adapt and incorporate technologies with their practices are heterogeneous. There was substantial heterogeneity in patterns of appropriation across individuals with respect to such things as approaches to email management, rhythms of use and number of messages sent and received. As Mackay (1988) found in her study more than 20 years ago, the 'use of electronic mail is strikingly diverse, although not infinitely so' (p. 344).

Teleology

A multilevel examination of teleology was undertaken providing an assessment of individual and organisational goals associated with use of email (Jasperson et al., 2005; van den Hooff, 2005). During the data collection for this case, DSTO made investments in two technologies designed to better support users in managing emails (Enterprise Vault) and to communicate via computer more dynamically (MS Communicator, which provides instant messaging). Enterprise Vault was introduced to assist DSTO in meeting its archiving responsibilities and to support staff in archiving and managing their messages. MS Communicator was deployed in an effort to provide staff with modern office communication technologies. These investments clearly influenced some users' patterns of appropriation, like an interviewee who had moved to using MS Communicator to support social communication needs in the workplace.

A range of belief and attitudinal influences, such as usability and usefulness, was assessed in this case. None of these influences was significantly related with any of the measures of appropriation. This is perhaps due to much of email users' appropriations not being goal directed but determined by habitual routines, such as rhythms of use, as well as by contextual influences, such as the nature of a job. This finding is consistent with the proposition put forward by Venkatesh et al. (2003) that such influences might be less likely to apply where behaviour is generated by routinised responses, rather than resulting from deliberate cognitions. Nevertheless, there was widespread evidence from the qualitative data that many of the participants had acted intentionally when changing their patterns of appropriation. The reasons given for making intentional changes included reflection on extant practices associated with email use, the introduction of other technologies and the patterns of use and suggestions of other people.

A teleological perspective draws attention to the purposeful pursuit of goals. Prior research has found that email is not only 'strikingly diverse' but that it also supports multiple goals (Wattenberg et al., 2005). In DSTO, email was used to support a wide variety of goals such as communication, information management and information sharing.

Dialectics

Dialectic process theory explains stability and change as the outcome of tensions between opposing or competing entities (Van de Ven and Poole, 1995). Tensions were analysed at the level of the email artefact and at the level of users' practices and associated technologies. Interviewees were asked to generate email similes. Ambivalence towards email was a prominent feature of these similes. For example, it is a 'necessary evil' that supports information exchange (thesis) but which also has a big impact on time (antithesis). Ambivalence represented the emotional synthesis resulting from tensions between the affordances (thesis) and constraints (antithesis) of email.

As has been discussed, email supports a variety of goals associated with communications, information management and information sharing (the thesis). There are other channels and technologies that also support the fulfilment of these goals, particularly face-to-face and telephone (the antithesis). The use of email, face-to-face, phone and other means of communication therefore provided users with a portfolio of channels and technologies to support the fulfilment of these goals (the synthesis).

Two technologies associated with communications and information management were introduced during the data collection phase of this case: Enterprise Vault and MS Communicator. These technologies represented the antithesis to the existing portfolio of technologies and practices associated with email (the thesis). Enterprise Vault was deployed to all desktops, with users having the option of installing MS Communicator if they so desired. Some individuals embraced these new technologies by incorporating them into their portfolios, as well as by adjusting their practices. The incorporation of the technologies with existing portfolios represented a synthesis. Also apparent were substitutions in functionality or affordances. Activities like informal communications and archiving that were previously undertaken using email were now performed using these new technologies. For example, one individual moved all of his archived messages back into his inbox so that they would all be located in the one place, with archiving of the inbox now done automatically by Enterprise

Vault, which in turn facilitated easier searching. Another individual had substituted email with MS Communicator to support informal communication and coordination amongst a walking group. The majority of other participants had not changed the location of their folders to their inbox to take advantage of the capabilities of Enterprise Vault, instead maintaining folders on their local PC or on a shared drive. They also had not installed MS Communicator.

Evolution

Evolution entails a continuous cycle of variation, selection and retention. Processes of variation and selection are associated with adaptations; retention is associated with stabilisations. All of these processes were manifested in this case.

Variations result from unpredictable events. Four interviewees identified accidents and incidents as influences on changes to their patterns of appropriation. One individual had previously kept all deleted items just in case he needed them, but accidentally deleted them one day and 'the sky didn't fall in'. From this point on, he decided to be 'a little more ruthless in culling things'. Similarly, another person had accidentally bulk deleted the messages in her inbox with no consequences. She subsequently decided to consciously bulk delete messages. Both of these accidents can be seen to have led to more efficient use of their time resources, the result of which was the selection of new email deletion practices.

Participants' patterns of appropriation were typified by both adaptation and inertia (retention). Participants retained similar patterns over time for checking and filing messages, messages sent, inbox size, number of folders and the type of email management approach adopted (inbox centric, folder centric or a combination). Furthermore, the majority of participants perceived their use of email as habitual.

The MTA Enhanced

Why did users' appropriations of email vary? From the perspective of the MTA, adaptations and stabilisations of patterns of appropriations were found, as was variability across individuals. All of these features are found in the MTA, and the sequencing of phases—adaptation then stabilisation then renewed adaptation—was also manifested. This provides a limited answer, however, to the 'why' question. An enhanced understanding of the process of appropriation was provided by examining the results from the perspective of three additional theories of change: teleology, dialectic and evolution.

- A multilevel view of intentionality assisted in providing a richer assessment of the rationale for changes in users' patterns of appropriation. Belief and attitudinal measures did not significantly influence users' appropriations, perhaps in part due to the lack of intentionality associated with various routinised patterns of appropriations. Nevertheless, many of the changes in patterns were intentional and were explained by reference to a variety of context-specific influences.
- Ambivalence was an important emotional synthesis that emerged from the tensions between the affordances and constraints of email. Email formed part of users' portfolios of communication channels and associated technologies that included phone, face-to-face and a variety of other channels. These channels and technologies were synthesised by users to create their portfolios. The process of synthesising new technologies into users' portfolios was also accompanied by substitution of affordances from the old onto the new.
- Accidents and incidents served as important influences on changes to patterns of appropriation for some users. Furthermore, the translation of the accidents into the selection of alternative appropriations was facilitated by the unintended efficiency gains derived from the accidents.

Conclusion

In this chapter the life cycle of IT use has been briefly outlined. A range of theories used to explain use has been considered and a candidate theory well suited to explaining the life cycle of use has been selected: the MTA. The MTA covers the entire life cycle from pre-use through to stablised use, as well as incorporating many of the strengths of the other theories and models. The MTA, however, is somewhat weak with respect to explaining the how and why of changes in patterns of use over the life cycle. The MTA largely relies on explanation by reference to an immanent logic as users move from one phase to another. For example, in the email case, adaptation was followed by stabilisation. To enhance explanation of the IT use life cycle, additional theories of change need to be incorporated. A teleological lens assists in explaining organisational and individual intentions associated with changes, such as the introduction of new technologies associated with email, as well as explaining how contextual influences contribute to changes in users' intentions and behaviours. The dialectic lens provides a way of explaining the substitutions, syntheses and maintenance of the status quo associated with certain patterns of appropriation over time. Finally, the evolutionary lens helps to explain how accidents and incidents play a role in shaping user behaviours and choices in the context of an environment constrained by limited time resources. To paraphrase Van de Ven and Poole (1995), it is the integration of these additional theories of change with the MTA that has led to the development of a new theory—the enhanced

MTA—which has stronger and broader explanatory power than the original MTA. In so doing, the enhanced MTA represents an important contribution to theorising about the life cycle of IT use.

References

Abraham, C., & Sheeran, P. (2004). Implications of goal theories for the theories of reasoned action and planned behavior. In C. Armitage & J. Christian (eds), *Planned Behavior: The relationship between human thought and action* (pp. 101–22). New Brunswick, NJ: Transaction Publishers.

Adams, D. A., Nelson, R. R., & Todd, P. A. (1992). Perceived usefulness, ease of use, and usage of information technology: a replication. *MIS Quarterly*, 16(2), 227–47.

Agarwal, R., & Prasad, J. (1998). The antecedents and consequents of user perceptions in information technology adoption. *Decision Support Systems, 22*(1), 15–29.

Ajzen, I. (1985). From intentions to actions: a theory of planned behavior. In J. Kuhl & J. Beckmann (eds), *Action Control: From cognition to behavior* (pp. 11–39). New York, NY: Springer Verlag.

Al-Gahtani, S., & King, M. (1999). Attitudes, satisfaction and usage: factors contributing to each in the acceptance of information technology. *Behaviour and Information Technology,18*(4), 277–97.

Bagozzi, R. P. (2007). The legacy of the technology acceptance model and a proposal for a paradigm shift. *Journal of the Association for Information Systems, 8*(4), 244–54.

Baker, E. (2007). Technology and its discontents. *Strategy+Business,* 31 July. Retrieved from <http://www.strategy-business.com>

Barley, S. R. (1986). Technology as an occasion for structuring: evidence from observations of CT scanners and the social order of radiology departments. *Administrative Science Quarterly, 31*(1), 78–108.

Baron, S., Patterson, A., & Harris, K. (2006). Beyond technology acceptance: understanding consumer practice. *International Journal of Service Industry Management, 17*(2), 111–35.

Benbasat, I., & Barki, H. (2007). Quo vadis, TAM? *Journal of the Association for Information Systems, 8*(4), 211–18.

Benbasat, I., & Zmud, R. W. (2003). The identity crisis within the IS discipline: defining and communicating the discipline's core properties. *MIS Quarterly, 27*(2), 183–94.

Bhattacherjee, A., & Premkumar, G. (2004). Understanding changes in belief and attitude toward information technology usage: a theoretical model and longitudinal test. *MIS Quarterly, 28*(2), 229–54.

Boersma, K., & Kingma, S. (2005). From means to ends: the transformation of ERP in a manufacturing company. *The Journal of Strategic Information Systems, 14*(2), 197–219.

Burton-Jones, A. & Straub, D. W., jr (2006). Reconceptualizing system usage: an approach and empirical test. *Information Systems Research, 17*(3), 228–46.

Carayannis, E. G., & Turner, E. (2006). Innovation diffusion and technology acceptance: the case of PKI technology. *Technovation, 26*(7), 847–55.

Carroll, J. (2004). Completing design in use: closing the appropriation cycle. Paper presented at the Twelfth European Conference on Information Systems (ECIS 2004), Turku, Finland.

Carroll, J., Howard, S., Murphy, J., & Peck, J. (2002a). 'No' to a free mobile: when adoption is not enough. *Proceedings of the 13th Australasian Conference on Information Systems (ACIS 2002)*, Melbourne, Australia.

Carroll, J., Howard, S., Peck, J., & Murphy, J. (2002b). A field study of perceptions and use of mobile telephones by 16 to 22 year olds. *Journal of Information Technology Theory and Application, 4*(2), 49–61.

Carroll, J., Howard, S., Peck, J., & Murphy, J. (2003a). From adoption to use: the process of appropriating a mobile phone. *Australasian Journal of Information Systems, 10*(2), 38–48

Carroll, J., Howard, S., Vetere, F., Peck, J., & Murphy, J. (2002c). Just what do the youth of today want? Technology appropriation by young people. *Proceedings of the 35th Annual Hawaii International Conference on Systems Sciences (HICSS 2002)*, 1777–85.

Carroll, J., Kriss, S., & Murphy, J. (2003b). Developing CRM systems that encourage user uptake. *Proceedings of the 7th Customer Contact World Conference*, Sydney, Australia.

Carroll, J., Mendoza, A., & Stern, L. (2005). Adoption, adaptation, stabilization and stagnation: software appropriation over time. *Proceedings of the 16th Australasian Conference on Information Systems (ACIS 2005)*, Sydney, Australia.

Chin, W. W., Gopal, A., & Salisbury, W. D. (1997). Advancing the theory of adaptive structuration: the development of a scale to measure faithfulness of appropriation. *Information Systems Research, 8*(4), 342–67.

Cho, S., Mathiassen, L., & Robey, D. (2007). Dialectics of resilience: a multi-level analysis of a telehealth innovation. *Journal of Information Technology, 22*(1), 24–35.

Chu, T.-H., & Robey, D. (2008). Explaining changes in learning and work practice following the adoption of online learning: a human agency perspective. *European Journal of Information Systems, 17*(1), 79–98.

Constantinides, P., & Barrett, M. (2006). Negotiating ICT development and use: the case of a telemedicine system in the healthcare region of Crete. *Information and Organization, 16*(1), 27–55.

Cooper, R. B., & Zmud, R. W. (1990). Information technology implementation research: a technological diffusion approach. *Management Science, 36*(2), 123–39.

Davis, F. D. (1989). Perceived usefulness, perceived ease of use, and user acceptance of information technology. *MIS Quarterly, 13*(3), 319–40.

Davis, F. D. (1993). User acceptance of information technology: system characteristics, user perceptions and behavioral impacts. *International Journal of Man-Machine Studies, 38*(3), 475–87.

Davis, F. D., Bagozzi, R. P., & Warshaw, P. R. (1989). User acceptance of computer technology: a comparison of two theoretical models. *Management Science, 35*(8), 982–1003.

DeLone, W. H., & McLean, E. R. (1992). Information systems success: the quest for the dependent variable. *Information Systems Research, 3*(1), 60–95.

DeSanctis, G., & Poole, M. S. (1994). Capturing the complexity in advanced technology use: adaptive structuration theory. *Organization Science, 5*(2), 121–47.

DeSanctis, G., Poole, M. S., & Dickson, G. W. (2000). Teams and technology: interactions over time. *Research on Managing Groups and Teams, 3*, 1–27.

Desouza, K. C., Awazu, Y., & Ramaprasad, A. (2007). Modifications and innovations to technology artifacts. *Technovation, 27*(4), 204–20.

Dishaw, M. T., & Strong, D. M. (1999). Extending the technology acceptance model with task-technology fit constructs. *Information and Management, 36*(1), 9–21.

Fidock, J. (2002). From evaluation to action: supporting the co-evolution of business practices and IT in the government sector. *Proceedings of the 9th European Conference on the Evaluation of IT (ECITE)*, Paris, France.

Giddens, A. (1986). *The Constitution of Society: Outline of the theory of structuration*. Berkeley, Calif.: University of California Press.

Goodhue, D. L. (2007). Comment on Benbasat and Barki's 'Quo vadis TAM' article. *Journal of the Association for Information Systems, 8*(4), 219–22.

Gopal, A., Bostrom, R. P., & Chin, W. W. (1992). Applying adaptive structuration theory to investigate the process of group support systems use. *Journal of Management Information Systems, 9*(3), 45–69.

Hage, J., & Aiken, M. (1970). *Social Change in Complex Organizations*. New York, NY: Random House.

Hayes, N. (1999). Work-arounds and boundary crossing in a high tech optronics company: the role of co-operative workflow technologies. *Computer Supported Cooperative Work, 9*(3–4), 435–55.

Herszfeld, S., Carroll, J., & Howard, S. (2003). Job allocation by SMS: technology appropriation in the construction industry. *Proceedings of the 14th Australasian Conference on Information Systems (ACIS 2003)*, Perth, Australia.

Heung, R. (2002). An evaluation of the model of technology appropriation, Honours thesis, University of Melbourne, Melbourne, Australia. Retrieved from <http://wwwtest.dis.unimelb.edu.au/future/research/theses/2002/HEUNG%282002%29_ abstract.pdf>

Hirschheim, R. (2007). Introduction to the special issue on 'Quo vadis TAM—issues and reflections on technology acceptance research'. *Journal of the Association for Information Systems, 8*(4): 203–5.

Hovorka, D. S., Germonprez, M., & Larsen, K. R. (2008). Explanation in information systems. *Information Systems Journal, 18*(1), 23–43.

Hsieh, J., Rai, A., & Keil, M. (2008). Understanding digital inequality: comparing continued use behavioral models of the socio-economically advantaged and disadvantaged. *MIS Quarterly, 32*(1), 97–126.

Hubona, G., & Geitz, S. (1997). External variables, beliefs, attitudes and information technology usage behavior. *Proceedings of the 30th Annual Hawaii International Conference on System Sciences (HICSS 97), Wailea, HI, 3*, 21–8.

Igbaria, M. (1990). End-user computing effectiveness: a structural equation model. *Omega, 18*(6), 637–52.

Igbaria, M., Pavri, F. N., & Huff, S. L. (1989). Microcomputer applications: an empirical look at usage. *Information and Management, 16*(4), 187–96.

Igbaria, M., Zinatelli, N., Cragg, P., & Cavaye, A. L. M. (1997). Personal computing acceptance factors in small firms: a structural equation model. *MIS Quarterly, 21*(3), 279–305.

Jasperson, J., Carter, P. E., & Zmud, R. W. (2005). A comprehensive conceptualization of post-adoptive behaviors associated with information technology enabled work systems. *MIS Quarterly, 29*(3), 525–57.

Johnson, B. M., & Rice, R. E. (1984). Reinvention in the innovation process: the case of word processing. In R. E. Rice (ed.), *The New Media: Communication, research, and technology* (pp. 157–83). Beverly Hills, Calif.: Sage.

Jones, M. R., & Karsten, H. (2008). Giddens's structuration theory and information systems research. *MIS Quarterly, 32*(1), 127–57.

Karahanna, E., Straub, D. W., & Chervany, N. L. (1999). Information technology adoption across time: a cross-sectional comparison of pre-adoption and post-adoption beliefs. *MIS Quarterly, 23*(2), 183–213.

Kautz, K., & Larsen, E. A. (2000). Diffusion theory and practice: disseminating quality management and software process improvement innovations. *Information Technology & People, 13*(1), 11–26.

Kim, R. M., & Kaplan, S. M. (2006). Interpreting socio-technical co-evolution: applying complex adaptive systems to IS engagement. *Information Technology & People, 19*(1), 35–54.

King, N. (1990). Innovation at work: the research literature. In M. A. West & J. L. Farr (eds), *Innovation and Creativity at Work: Psychological and organizational strategies* (pp. 15–59). Oxford, UK: John Wiley & Sons.

Kraut, R. E., Rice, R. E., Cool, C., & Fish, R. S. (1998). Varieties of social influence: the role of utility and norms in the success of a new communication medium. *Organization Science, 9*(4), 437–53.

Kwon, T. H., & Zmud, R. W. (1987). Unifying the fragmented models of information systems implementation. In R. J. Boland & R. A. Hirschheim (eds), *Critical Issues in Information Systems Research* (pp. 227–51). New York, NY: John Wiley & Sons.

Lee, A. S. (1999). Researching MIS. In W. L. Curry & R. Galliers (eds), *Rethinking Management Information Systems* (pp. 7–27). Oxford, UK: Oxford University Press.

Leonard-Barton, D. (1988). Implementation as mutual adaptation of technology and organization. *Research Policy 17*(5), 251–67.

Lin, H.-F., & Lee, G.-G. (2006). Effects of socio-technical factors on organizational intention to encourage knowledge sharing. *Management Decision, 44*(1), 74–88.

Lucas, H. C., jr, Swanson, E. B., & Zmud, R. W. (2007). Implementation, innovation, and related themes over the years in information systems research. *Journal of the Association for Information Systems, 8*(4), 206–10.

Lyytinen, K., & Newman, M. (2008). Explaining information systems change: a punctuated socio-technical change model. *European Journal of Information Systems, 17*(6), 589–613.

Mackay, W. E. (1988). More than just a communication system: diversity in the use of electronic mail. *Proceedings of the 1988 ACM Conference on Computer-Supported Cooperative Work (CSCW 88)*, Portland, Ore., 344–53.

Mackay, W. E. (1990a). Patterns of sharing customizable software. *Proceedings of the 1990 ACM Conference on Computer-Supported Cooperative Work (CSCW 90)*, Los Angeles, Calif., 209–21.

Mackay, W. E. (1990b). Users and customizable software: a coadaptive phenomenon. Doctoral dissertation, Massachusetts Institute of Technology, Cambridge, Mass.

McLean, E., Seddon, P., Torkzadeh, R., & Wilcocks, L. (2002). IS success measurement: guidelines for future research. *Proceedings of the Twenty-Third International Conference on Information Systems (ICIS 2002)*.

McMaster, T., & Wastell, D. (2005). Diffusion—or delusion? Challenging an IS research tradition. *Information Technology & People, 18*(4), 383–404.

Majchrzak, A., Rice, R. E., Malhotra, A., King, N., & Ba, S. (2000). Technology adaptation: the case of a computer-supported inter-organizational virtual team. *MIS Quarterly, 24*(4), 569–600.

Markus, M. L., & Silver, M. S. (2008). A foundation for the study of IT effects: a new look at DeSanctis and Poole's concepts of structural features and spirit. *Journal of the Association for Information Systems, 9*(10), 609–32.

Mathieson, K. (1991). Predicting user intentions: comparing the technology acceptance model with the theory of planned behavior. *Information Systems Research, 2*(3), 173–91.

Mendoza, A., Carroll, J., & Stern, L. (2005). Adoption, adaptation, stabilization and stagnation: software appropriation over time. *Proceedings of the 16th Australasian Conference on Information Systems (ACIS 2005)*, Sydney, Australia.

Mendoza, A., Carroll, J., & Stern, L. (2008). Influences on continued use of an information system: a longitudinal study. *Proceedings of the 16th European Conference on Information Systems (ECIS)*, Galway, Ireland.

Mendoza, A., Stern, L., & Carroll, J. (2007). Plateaus in long-term appropriation of an information system. *Proceedings of the 18th Australasian Conference on Information Systems (ACIS 2007)*, Toowoomba, Australia.

Mintzberg, H., & Westley, F. (2001). Decision-making: it's not what you think. *MIT Sloan Management Review, 42*(3) 89–93.

Moore, G. C., & Benbasat, I. (1991). Development of an instrument to measure the perceptions of adopting an information technology innovation. *Information Systems Research, 2*(3), 192–222.

Myers, M. D. (1994). Dialectical hermeneutics: a theoretical framework for the implementation of information systems. *Information Systems Journal, 5*(1), 51–70.

Nord, W. R., & Tucker, S. (1987). *Implementing Routine and Radical Innovations*. San Francisco, Calif.: New Lexington Press.

Nor Zairah, A. R., & Rose Alinda, A. (2007). Open source software appropriation: attractor and repellent criteria of a city council. Paper presented at the MMU International Symposium on Information and Communications Technologies (M2USIC 2007).

Ogden, J. (2003). Some problems with social cognition models: a pragmatic and conceptual analysis. *Health Psychology, 22*(4), 424–8.

Orlikowski, W. J. (1992). The duality of technology: rethinking the concept of technology in organizations. *Organization Science, 3*(3), 398–427.

Pfeffer, J. (1982). *Organizations and Organization Theory*. Marshfield, Mass.: Pitman.

Pollard, C. (2003). Exploring continued and discontinued use of IT: a case study of OptionFinder, a group support system. *Group Decision and Negotiation, 12*(3), 171–93.

Poole, M. S., & DeSanctis, G. (1990). Understanding the use of group decision support systems: the theory of adaptive structuration. In J. Fulk & C. Steinfield (eds), *Organizations and Communication Technology* (pp. 173–93). Beverly Hills, Calif.: Sage.

Pozzebon, M. (2004). The influence of a structurationist view on strategic management research. *Journal of Management Studies, 41*(2), 247–72.

Prescott, M. B., & Conger, S. A. (1995). Information technology innovations: a classification by IT locus of impact and research approach. *Data Base Advances, 26*(2–3), 20–41.

Raymond, L. (1985). Organizational characteristics and MIS success in the context of small business. *MIS Quarterly, 9*(1), 37–52.

Rice, R. E., & Rogers, E. M. (1980). Reinvention in the innovation process. *Science Communication, 1*(4), 499–514.

Roberts, P., & Henderson, R. (2000). Information technology acceptance in a sample of government employees: a test of the technology acceptance model. *Interacting with Computers, 12*(5), 427–43.

Robey, D., & Boudreau, M.-C. (1999). Accounting for the contradictory organizational consequences of information technology: theoretical directions and methodological implications. *Information Systems Research, 10*(2), 167–85.

Robey, D., Ross, J. W., & Boudreau, M.-C. (2002). Learning to implement enterprise systems: an exploratory study of the dialectics of change. *Journal of Management Information Systems, 19*(1), 17–46.

Rogers, E. M. (1995). *Diffusion of Innovations*. New York, NY: Free Press.

Rose, J., & Jones, M. (2005). The double dance of agency: a socio-theoretic account of how machines and humans interact. *Systems, Signs & Actions, 1*(1), 19–37.

Rosemann, M., & Vessey, I. (2008). Toward improving the relevance of information systems research to practice: the role of applicability checks. *MIS Quarterly, 32*(1), 1–22.

Sabherwal, R., Hirschheim, R., & Goles, T. (2001). The dynamics of alignment: insights from a punctuated equilibrium model. *Organization Science, 12*(2), 179–97.

Saga, V. L., & Zmud, R. W. (1994). The nature and determinants of IT acceptance, routinization and infusion. *Proceedings of the IFIP TC8 Working Conference on Diffusion, Transfer and Implementation of Information Technology*, 67–86.

Salisbury, D., Chin, W. W., Gopal, A., & Newsted, P. R. (2002). Research report: better theory through measurement—developing a scale to capture consensus on appropriation. *Information Systems Research, 13*(1), 91–103.

Schwarz, A., & Chin, W. (2007). Looking forward: toward an understanding of the nature and definition of IT acceptance. *Journal of the Association for Information Systems, 8*(4), 230–43.

Sharma, R., & Yetton, P. (2001). An evaluation of a major validity threat to the technology acceptance model [research in progress]. *Proceedings of the 9th European Conference on Information Systems*, Bled, Slovenia, 1170–5.

Shih, H.-P. (2008). Continued use of a Chinese online portal: an empirical study. *Behaviour and Information Technology, 27*(3), 201–9.

Silva, L. (2007). Post-positivist review of technology acceptance model. *Journal of the Association for Information Systems, 8*(4), 255–66.

Straub, D. W., jr, & Burton-Jones, A. (2007). Veni, vidi, vici: breaking the TAM logjam. *Journal of the Association for Information Systems, 8*(4), 223–9.

Sundaram, S., Schwarz, A., Jones, E., & Chin, W. W. (2007). Technology use on the front line: how information technology enhances individual performance. *Journal of the Academy of Marketing Science, 35*(1), 101–12.

Taylor, S., & Todd, P. A. (1995a). Assessing IT usage: the role of prior experience. *MIS Quarterly, 19*(4), 561–70.

Taylor, S., & Todd, P. A. (1995b). Understanding information technology usage: a test of competing models. *Information Systems Research, 6*(2), 144–76.

Trice, A. W., & Treacy, M. E. (1988). Utilization as a dependent variable in MIS research. *Data Base, 19*(3–4), 33–41.

Trigg, R. H., & Bødker, S. (1994). From implementation to design: tailoring and the emergence of systematization in CSCW. *Proceedings of the 1994 ACM Conference on Computer Supported Cooperative Work (CSCW 94)*, 45–54.

Tyre, M. J., & Orlikowski, W. J. (1994). Windows of opportunity: temporal patterns of technological adaptation in organizations. *Organization Science, 5*(1), 98–118.

van den Hooff, B. (2005). A learning process in email use—a longitudinal case study of the interaction between organization and technology. *Behaviour & Information Technology, 24*(2), 131–45.

Van de Ven, A., & Poole, M. S. (1995). Explaining development and change in organizations. *Academy of Management Review, 20*(3), 510–40.

Van de Ven, A., & Poole, M. S. (2005). Alternative approaches for studying organizational change. *Organization Studies 26*(9), 1377–404.

Venkatesh, V., & Bala, H. (2008). Technology acceptance model 3 and a research agenda on interventions. *Decision Sciences, 39*(2), 273–315.

Venkatesh, V., Morris, M. G., Davis, G. B., & Davis, F. D. (2003). User acceptance of information technology: toward a unified view. *MIS Quarterly, 27*(3), 425–78.

Wattenberg, M., Rohall, S. L., Gruen, D., & Kerr, B. (2005). E-mail research: targeting the enterprise. *Human-Computer Interaction, 20*(1–2), 139–62.

Wolfe, R. A. (1994). Organizational innovation: review, critique and suggested research directions. *Journal of Management Studies, 31*(3), 405–31.

5. A Critical Systems Thinking Perspective for IS Adoption

Syed Arshad Raza
Edith Cowan University

Craig Standing
Edith Cowan University

Abstract

Information systems (IS) project management is a challenging task. Lack of user support and involvement are among the key reasons for IS/IT project failure. The established information technology adoption models—like TAM, TAM2 and similar—only consider technology adoption from an individual user's viewpoint, highlighting key factors and their relationships, but they do not provide any mechanism to deal with multiple user perspectives and their roles in a holistic framework from a project management viewpoint. This chapter proposes a model for information system adoption based on critical systems thinking (CST) in an organisational context from a management perspective. The authors, considering IS adoption as a multiphase innovation project, argue that boundary considerations using a multiple stakeholder perspective (boundary critique) provide an alternative focus for IS adoption. The chapter uses the five basic phases or activities for information system development of the system development life cycle (SDLC). The model integrates the traditional SDLC with the ongoing process of 'phase-stakeholder-identification'. The emerging systemic stakeholder networks are proposed to be applied with network mechanisms to influence stakeholders' attitudes towards IS adoption. The study suggests that the proposed model has the capacity to serve as a roadmap for smoother IS adoption by facilitating organisational learning and change.

Introduction

Failures in information systems (IS) or information technology (IT) projects quite frequently occur, indicating the challenging nature of the IS/IT project management task (Azzara and Garone, 2003; Chen and Latendresse, 2003). Standing et al. (2006), among many others, have identified the major reasons for such failures as the lack of user support and involvement, lack of support and commitment of executive management, imprecisely defined project objectives and poor project management and leadership.

The application and use of IT in organisations have been extensively researched over the past few decades. The technology acceptance model, or TAM (Davis, 1989; Davis et al., 1989), has significantly contributed to the organisational know-how related to user acceptance of technology. TAM was later extended to TAM2 (Venkatesh and Davis, 2000) and the unified theory of acceptance and use of technology, or UTAUT (Venkatesh et al., 2003), but despite providing an insight into the key factors and their relationships that influence user acceptance of IT in organisations, these models do not provide any mechanism showing how an organisation can successfully proceed in IS adoption by taking multiple stakeholder perspectives and their roles of involvement into account. On the other hand, IS methodologies like the waterfall, prototyping and evolutionary models fall short of addressing issues of perception, expectancy, internal or external politics and cognitive processes that can result in IS project failure (Yardley, 2002).

This chapter aims to provide a methodological model for IS adoption in an organisational context from a critical systems thinking perspective. We consider IS adoption using Ulrich's notion of boundary considerations (boundary critique), which involves multiple stakeholders, because we believe that it can effectively help in addressing the challenges of IS adoption and provide for smoother organisational learning and change. Information systems adoption is regarded here as a multiphase innovation project, and an information system adoption model using the phases of systems development life cycle (SDLC) is proposed as an example and our model is based on concepts related to network stakeholder theory, critical systems thinking (CST) and innovation diffusion. It should be noted that the chapter uses the terms IS and IT interchangeably.

The chapter comprises four main sections. The first section analyses the literature and highlights the concepts that underpin the proposed model; the second presents the proposed model; the third discusses its implications; and the last section focuses on limitations and overall conclusions.

Literature Analysis

A Management Perspective of the Stakeholder Theory

Stakeholders are a consistent presence in any organisational life cycle (Rowley 1997). It was Freeman who brought stakeholder theory into the mainstream of management literature, defining a stakeholder as any group or individual who can affect or is affected by the achievement of the firm's objectives (Freeman, 1984). He conceptualised the firm or the focal organisation (FO) as the hub of a wheel and stakeholders as the ends of spokes around it, as depicted in Figure 5.1.

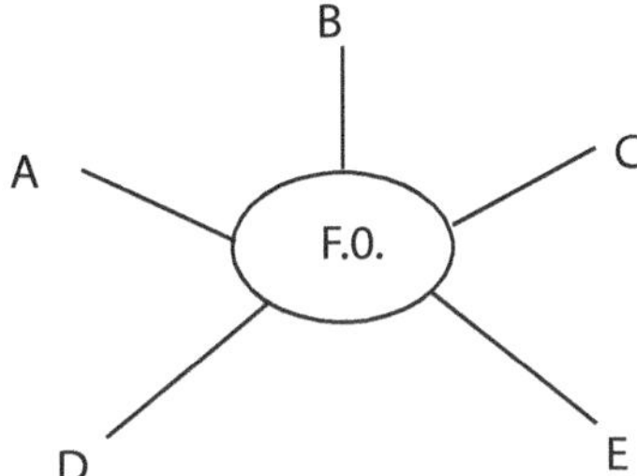

Figure 5.1 Freeman's Hub-and-Spoke Model

But Freeman's (1984) hub-and-spoke model, as mentioned by Rowley (1997), could not portray a realistic picture because:

1. it is highly likely that direct relationships among stakeholders exist, which means there is in fact a network of stakeholders and their influences
2. firms do not simply respond to each stakeholder individually; rather, they respond to multiple influences from the entire stakeholder set or a group of stakeholders
3. the organisation does not necessarily lie at the centre of the network, but is rather a stakeholder in its relevant social system or network of stakeholders.

The original model was therefore extended by Freeman and Evan (1990), emerging as a series of multilateral contracts among stakeholders and giving birth to a network of stakeholders, as shown in Figure 5.2. Thus, explaining an organisation's response to its stakeholders requires an analysis of a complex and interdependent array of relationships among stakeholders and their roles rather than just their individual relationships with the organisation.

This refined and extended view of stakeholders by Freeman and Evan (1990) forms the basis of the network of stakeholders, which we call the 'systemic stakeholder network' for our proposed model, as shown in Figure 5.5. This network involves stakeholders and the roles they play during various IS development phases.

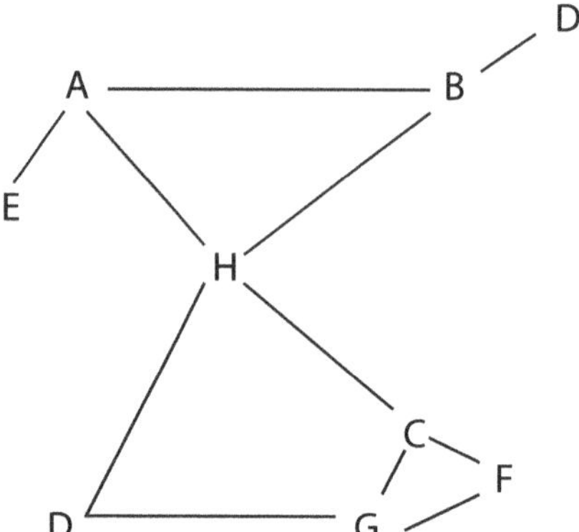

Figure 5.2 Network of Stakeholders

Systems Thinking and Critical Systems Thinking (CST)

Systems or *holistic thinking* views a system as a whole, focusing on how its parts affect the whole through their interactions (Ackoff, 1995) and studying their multiple cross-level interactions over time (Waldman, 2007). Critical systems thinking (CST), proposed by Ulrich (1983), is a systems thinking-based framework for a reflective practice that considers a social system design by defining its boundary as those who are *involved* in and *affected* by it. It moved Churchman's (1970, 1979) understanding of systemic boundary analysis in a new and challenging direction termed 'boundary critique' (Midgley, 2007). This chapter applies this approach to IS adoption in an organisational context.

The concept of 'boundary critique' is based on the idea of whole systems improvement (Achterkamp and Vos, 2007). It aims to include or *sweep-in* the maximum amount of information into the defined system boundary for analysis, on one hand, and poses the question of the rational justification of this boundary through a debate between stakeholders on the other, thus making it an ethical process involving multiple viewpoints. A different system boundary might result in a different problem analysis and, accordingly, in different solutions or changes. Wilby (2005) argues that the goal of holistic study is not to *sweep-in* or include everything involved; rather, it is about deciding what is relevant to the study and what is not and understanding the reasons for those choices. The choices are affected by biases and interests about what is likely to be included in the system and what is considered to be the system's environment. For example, if a car, producing the desired level of power output, is causing environmental pollution because of the unhealthy composition of its emissions then *sweeping-in* the environmental safety consideration into the boundary of analysis might lead to an entirely opposite system evaluation outcome.

Ulrich (1983) provides practical guidelines that both planners and ordinary citizens can use equally proficiently to conduct boundary critique. For this purpose, he offers a list of 12 questions that can be employed by those involved in and affected by the working out of what the system currently *is* and what it *ought* to be. The boundary consideration thus coherently defines *what* issues are to be included or excluded and *who is to be involved* (the stakeholders) with these issues (Midgley, 2003).

Technology in organisations, according to Malmsjö and Övelius (2003), is related to systems that use specific methods to process matter or information. This results in products that satisfy the needs of different stakeholders in society. It is, therefore, quite critical to identify stakeholders and discover how IS adoption is looked at through the eyes of those involved in and/or affected by it. From an IS adoption perspective, those involved can have an influence on the achievement of the objectives of the adoption process whereas those affected are influenced by the achievement of these objectives.

Change Management and Innovation Diffusion

Through a detailed literature analysis, Malmsjö and Övelius (2003) have identified change factors that influence IS in organisations. They have classified these into 'internal' and 'external' factors. Hall and Hord (2006), however, indicate that the success of a change process depends less on whether the source of change is internal or external. Rather, it depends more on the degree of openness and readiness of an organisation to consider the actions being undertaken and continually to examine ways to learn.

Hall and Hord (2006) view change as an innovation diffusion process rather than an event. The dynamic nature of IS necessitates that organisations and researchers understand and manage diffusion of innovations (Nilakanta and Scamell, 1990). Innovation is defined as an idea, a practice or a material artefact (product) such as a computer that is perceived to be new by the relevant unit of adoption (Hall and Hord, 2006; Zaltman et al., 1973). Innovation diffusion, according to Rogers (1995), is the process by which an innovation is communicated among the members of a social system through certain channels over time.

Innovation diffusion is often a victim of poor change management as organisations fail to realise that the resistance offered by people is not necessarily to the change per se, but the way they are treated and the roles they play in the change process (Cooper and Markus, 1995). Organisational participants who are vaguely aware of the process can cause rumours and anxiety resulting in attitudes different from those intended by management and which ultimately lead to resistance (Jick, 1993).

Our proposed model identifies stakeholders and their roles over time, and uses the emerging stakeholder network for diffusing information about the IS project to influence attitudes. The details of the model are given in the next section.

A Proposed Model for IS Adoption

This section first formulates the basic principles, based on the literature, needed to lay the foundations of the IS adoption model. It then presents the model itself by integrating the systems development life cycle (SDLC) with concepts related to critical systems thinking (CST), network stakeholder theory and innovation diffusion.

Grounded upon the reviewed literature, we formulate the following two basic principles.

Principle 1

Information systems adoption is a multiphase innovation project, consisting of a series of steps viewed as change processes, not events (Hall and Hord, 2006; Rogers, 1995).

Principle 2

Change is by definition a dynamic process (Cao et al., 2003) that makes identification of stakeholders and their roles an ongoing process based on organisational learning, often resulting in the redefinition of boundaries of the system as the IS adoption process progresses.

Composition of the Proposed Model

This subsection discusses the components that constitute our proposed IS adoption model.

Methodology

Unlike software, information systems are never *off-the-shelf* as they revolve around a company's people (stakeholders) and procedures (Kroenke, 2009). As per principle one, IS adoption is viewed as a purposeful innovation project and the diffusion of an innovation is viewed as a phased process (Rogers, 1995). The phases underpinning the process of information system development (ISD) vary radically depending on the chosen methodology. There are, however, five basic activities or phases that are shared—albeit with different names—by most methodologies. These are: 1) identification and concept; 2) requirements definition; 3) system analysis and design; 4) implementation; and 5) testing and operation (Carugati, 2008).

The proposed model considers these activities of the SDLC methodology as examples of IS adoption in an organisational context. The SDLC is a traditional systems development methodology (see Figure 5.3) with a well-defined process for conceiving, developing and implementing an information system (Mahmood, 1987). Figure 5.3 illustrates these activities (with different names) carried out at each stage of the SDLC. It also highlights their relationship and interdependence. There are, however, problems of systems delivery and communication pertaining to the SDLC (Berrisford and Wetherbe, 1979; Gremillion and Pyburn, 1983) that we will address later.

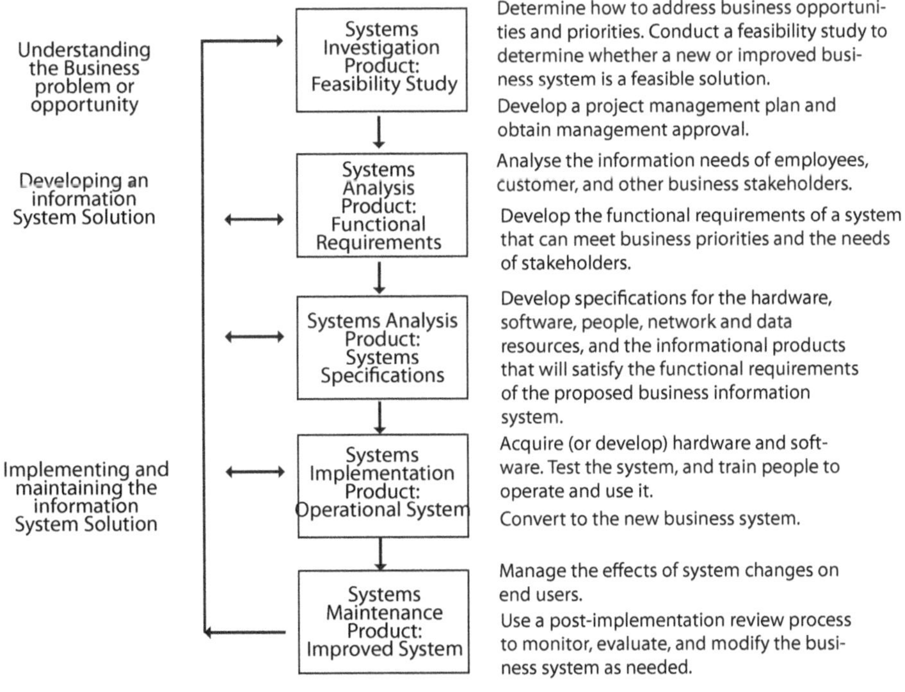

Figure 5.3 The Traditional IS Development Cycle

Source: O'Brien and Marakas (2005, p. 343).

We have adopted the SDLC as a roadmap for IS adoption progressing through its various phases, as shown in Figure 5.4. Figure 5.4 also shows that, on the basis of learning, IS project activities may be recycled back at any time to repeat previous activities with the aim of modifying and improving the system being developed (O'Brien and Marakas, 2005).

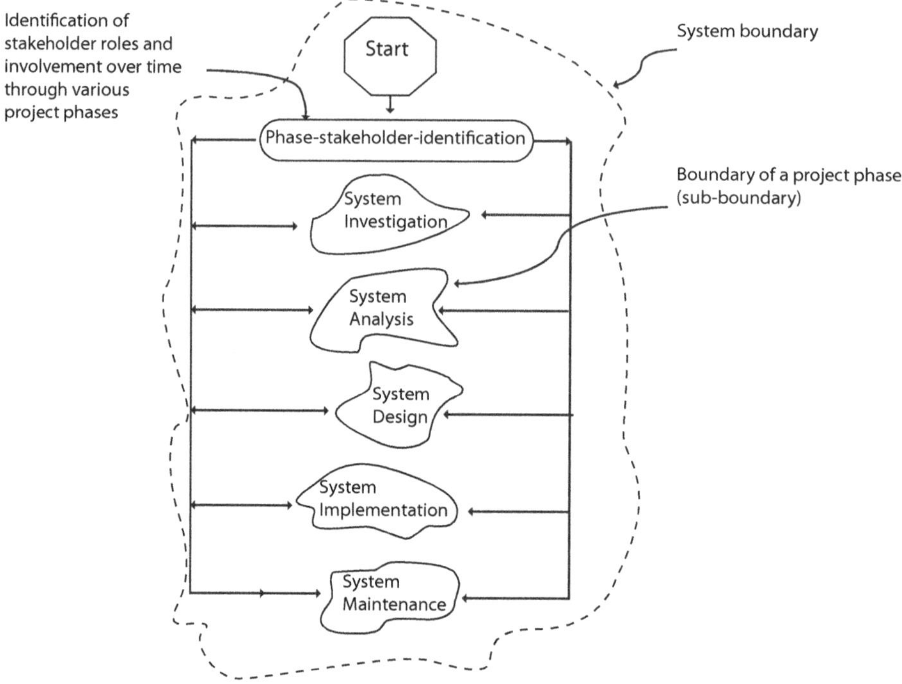

Figure 5.4 The Proposed Critical Systems Thinking-Based IS Adoption Model Using SDLC Phases

Phase-Stakeholder-Identification Using Boundary Critique

Identifying stakeholders, in the view of Vos (2003), is to draw a line between the parties to be involved and the parties not to be involved. Achterkamp and Vos (2007) propose a four-step method for project-based stakeholder identification using boundary critique that focuses on two key points: *roles of involvement* and *phasing this involvement*. They define a project broadly as an innovation project specially set up for pursuing the development of new products, services or processes (IS adoption, for instance), or a project concerning a (temporary) task inside or outside an organisation. The *roles of involvement* are underpinned by Ulrich's (1983) notion of boundary critique while *phasing of involvement* relates these roles to the dynamic processes of a project encompassing four phases of *initiation, development, implementation* and *maintenance*.

Based on the stakeholder roles defined by Achterkamp and Vos (2007), Table 5.1 defines the roles that stakeholders play in the context of the proposed model. The roles of stakeholders listed in the table fall into two main categories—those involved and those affected (termed actively and passively involved respectively)—while the other roles might fall into either of these categories. In Figure 5.5, stakeholders W, X, Y, Z and, in Figure 5.6, stakeholders S, T, U, W,

X, Y and Z have been shown to exemplify those actively involved or passively involved (affected) and not playing the roles of a client (C), a decision maker (DM) or a designer (D). Client C has been shown at the intersection of the actively involved and passively involved (affected) stakeholders though this may vary across different project scenarios.

Table 5.1 Roles of Involvement for IS Adoption

Role	Definition
Party involved actively and passively (the two basic categories)	A *party involved* is any group or individual who i. can affect the achievement of the project objectives (*actively involved*) or ii. is affected by the achievement of these objectives (*passively involved*).
Client (C)	A *client* is a party whose purposes are being served through the project.
Decision maker (DM)	A *decision maker* is responsible for i. identifying business opportunities and priorities in relation to the IS project ii. conducting a feasibility study about the new or improved IS iii. analysing the information needs of stakeholders iv. setting requirements regarding the project processes and outcomes and evaluating whether these requirements are met v. managing the effects of system changes on end users vi. establishing and/or revamping stakeholder networks for IS diffusion vii. monitoring and evaluating a post-implementation review.
Designer (D)	A *designer* contributes expertise within the IS project and is responsible for the i. (interim) deliverables ii. development of a project management plan and its approval iii. development of functional requirements that could meet the business priorities and the needs of stakeholders iv. development of specifications for the hardware, software, people, network and data resources v. system testing and user training vi. modifications to the IS based on a post-implementation review.
Passively involved representative (R)	A *passively involved* representative is affected by the project outcomes or project process without being able to influence the process or these outcomes. A *representative* is a person who has been chosen to act on behalf of another—that is, the passively involved.

Unlike the four project phases identified by Achterkamp and Vos (2007), we consider IS adoption under the five basic activities of ISD (Carugati, 2008) or project phases as defined under the SDLC—namely, investigation, analysis, design, development and maintenance (O'Brien and Marakas, 2005). Moreover, based on the four-step stakeholder identification method suggested by Achterkamp and Vos (2007), we emphasise the ongoing requirement for identification of stakeholders and its repetition as required during the progress of the IS adoption project (see *principle two*), as shown in Figure 5.4.

We term this *phase-stakeholder-identification* and it generates the *systemic network of stakeholders* (Figure 5.5) while its repetition *sweeps-in* more information based on the effectiveness of the strategies for innovation diffusion, applied in the previous cycle(s).

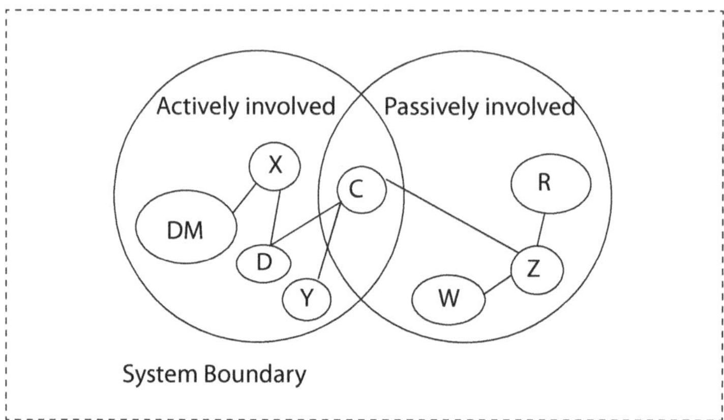

Figure 5.5 Systemic Network of Stakeholders

This repetition will eventually result in the redefinition of the boundaries under consideration, establishing the *network of stakeholders* as a function of time, as shown in Figure 5.6.

These networks emerging over time we call a *systemic network of stakeholders*. This model, based on the idea of *progressive boundary refinement*, consists of the following steps.

1. Define the goal of the project phase.
2. Identify stakeholders for each SDLC phase on the basis of *phases of involvement* (see Figure 5.6) and *roles of involvement* (see Table 5.1).
3. Represent stakeholders in the form of a *systemic network of stakeholders* over time (see Figures 5.5 and 5.6).
4. Apply network mechanisms (described in the next section) for influencing attitudes regarding IS adoption of the stakeholders in the network.

Since the SDLC phases have just been used to exemplify the five basic activities of ISD as identified by Carugati (2008), the process of *phase-stakeholder-identification* can also be coupled with methodologies other than the SDLC. The discussion here as to how this can be achieved has, however, been restricted due to space limitations.

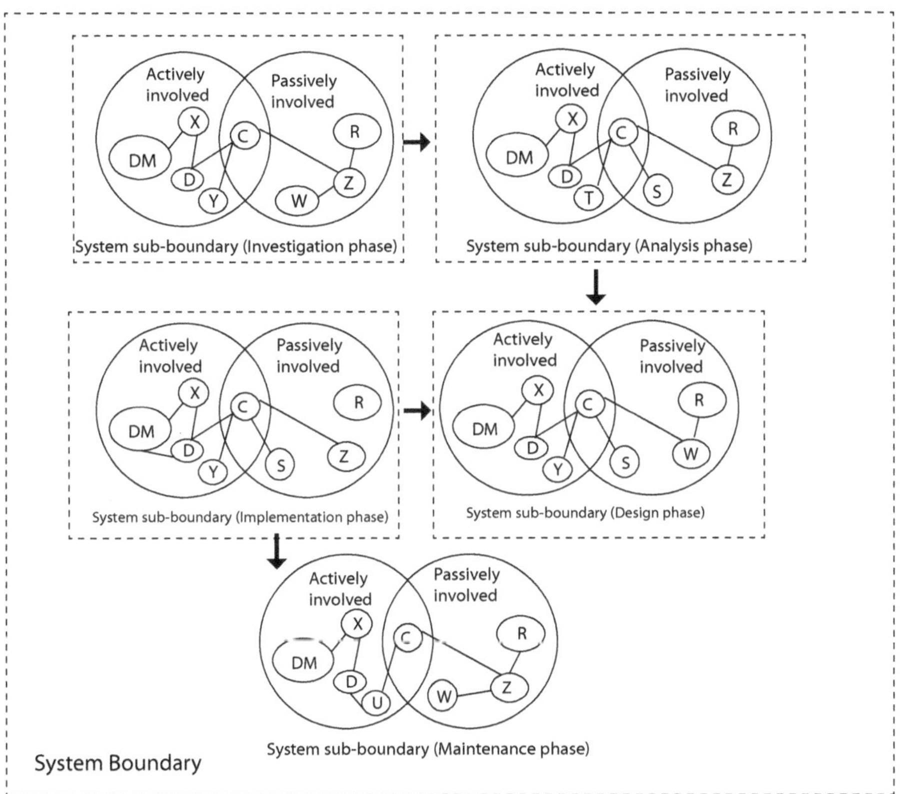

Figure 5.6 Emergence of Systemic Stakeholder Networks Over Time Through Various SDLC Phases

Communication of Innovation

Innovation diffusion, according to Rogers (1995), is the process by which an innovation is communicated among the members of a social system through certain channels over time. Accordingly, we now highlight the applicability of network mechanisms to the systemic network of stakeholders to influence opinions about the IS adoption process. The availability of information about the innovation and the communication processes heavily influences the diffusion process between the change proponents (the ones actively involved) and those who are affected (passively involved) by it (McIlduff and Coghlan, 2000).

As indicated by Cao et al. (2003), a shift in organisational form is tending to take place from rational bureaucratic composition towards a network-based configuration, characterised by a flatter authority structure. This configuration comprises multiple horizontal linkages between the inner core of a firm and its outside suppliers, contractors and customers (its stakeholders). This network of stakeholder relationships can be studied and analysed using social network analysis. This analysis has been used by researchers to refine and extend understanding of various behavioural and social phenomena, including community elite decision making, social influence, power and innovation diffusion (Cao et al., 2003; Rowley, 1997). According to Cummings and Worley (1993, p. 288), 'the network structure is highly flexible. Its components can be assembled and reassembled to meet changing conditions.'

For communicating information regarding IS adoption, we have emphasised the establishment of systemic networks of stakeholders (see Figures 5.5 and 5.6) over time by using 'boundary critique'. Network theorists argue that such networks influence perceptions and opinions and are capable of changing interpretations associated with and reducing uncertainty about an event, idea or phenomenon (Rogers and Kincaid, 1981). We argue that this capability of networks can be used in managing IS adoption and can influence perceptions of stakeholders about the process, and begin by providing a brief overview of the network mechanisms discussed in the literature.

Relational proximity or communication proximity views organisation as a communication network in which stakeholders repeatedly interact (directly and indirectly) to process resources and information (Dow, 1988, p. 56; Rogers and Kincaid, 1981). As mentioned by Erickson (1988), people are most likely to compare and agree with those to whom they are more strongly tied.

Positional proximity refers to the network of structurally proximate individuals who might not have links with one another as in relational proximity but they are linked to others with similar attributes like roles and obligations, status and expectations (Burt, 1980). 'Individuals may be the focus of similar information, requests and demands from members of their role set, creating an information field in which they are embedded' (Hartman and Johnson, 1989, p. 525).

Spatial proximity is based on the likelihood of interaction and exposure to social information due to living or working close together, which influences one's attitudes (Festinger et al., 1950). Unlike direct interaction, it can affect social information processing through exposure to or inaccessibility of the individuals to the organisational sub-climates, task materials and events (Hackman, 1983).

Together with these network mechanisms, we suggest the use of interventions for information propagation about IS adoption. In the context of innovation diffusion, an intervention is an action or event that influences (positively or negatively) the individuals involved or expected to be involved in the process (Hall and Hord, 1987, p. 143). McIlduff and Coghlan (2000, p. 724) view interventions as 'all conscious and deliberate actions and behaviours on the part of a manager, consultant or facilitator'. Our position is that network mechanisms combined with appropriate intervention strategies will influence the attitudes and behaviour of participants in conflict situations. The network mechanism will serve as a medium for information flow while the nature of intervention and the roles played by opinion leaders during these interventions will collectively determine the likelihood of innovation adoption success. Focusing on various intervention types is not, however, within the scope of this chapter. Hall and Hord (2006) provide a detailed discussion of various intervention sizes, functions, their levels and anatomy.

Discussion

Our model revolves around innovation diffusion in the context of IS adoption. Rogers (1995) considers members of the social system (people), communication channels and time to be the main ingredients of the innovation diffusion process. Azzara and Garone (2003) and Standing et al. (2006) regard the factors of the following list as key for the success of an IT/IS project

1. stakeholder support and involvement
2. project management and leadership
3. effective planning
4. executive commitment
5. project team commitment.

We now discuss how our model suggests these key factors might be addressed. The factors will be referred to in what follows as key factor one, key factor two, and so on.

As described above, our model has three main components: methodology, phase-stakeholder-identification and communication. Critical systems thinking (CST) is applicable in IS adoption due to its commitment to human/stakeholder involvement (key factor one) through the use of boundary critique. Moreover, IS adoption inside an organisation has impacts on its actions through its orientation to the roles and responsibilities of its stakeholders over time, and the adoption process will not prove successful until stakeholders and their roles are identified during the various phases governing IS adoption, based on the boundaries under consideration. The proposed model thus applies boundary critique over the time dimension, addressing key factor one.

Due to CST's handicap in providing guidance on issues like process re-engineering, product quality improvement and applications development (Cao et al., 2003), the systems development life cycle (SDLC) has been used here to engage the methodological component of the model. It, in fact, has been used to serve as a roadmap for the IS adoption process and provides guidance to critically examine the progress of the whole project and the decision-making process. The reason for opting for a methodology for the IS adoption process can be justified as it keeps people focused on the proper tasks and activities required at a specific step of a transformation project (Paper et al., 2000). It serves as a rallying venue for cross-functional teams, facilitators and managers by keeping them informed regarding project progress (Kettinger et al., 1997), addressing key factors two and three.

Since diffusion of innovation is affected by the sources of information and channels of communication (Nilakanta and Scamell, 1990), the third component of communication is taken care of by the continuous determination of the stakeholder interactions in the form of *systemic stakeholder networks* that emerge over time as the IS adoption progresses through the SDLC phases. Singh (2005) has, for example, empirically shown the effectiveness of collaborative networks in knowledge flow and its diffusion.

Our argument is that repetition of defining/redefining stakeholders and their roles and the application of the network mechanisms to *systemic stakeholder networks* have the potential to serve as a rudder for the IS adoption process throughout the project life cycle (see Figure 5.6), resulting in effective communication management and in overcoming problems of systems delivery and communication pertaining to the SDLC (Berrisford and Wetherbe, 1979; Gremillion and Pyburn, 1983). This approach would show the commitment and concern of opinion leaders or executives and the project teams about the success of the IS adoption project by keeping the stakeholders (actively involved and passively involved) well informed about the project objectives and progress, addressing key factors one, four and five.

As shown in Figure 5.4, the SDLC has been coupled with *phase-stakeholder-identification* prior to each project phase. The irregular shapes illustrated around project phases (systems investigation, analysis, design, implementation and maintenance) represent each phase as an amoeba—a microscopic organism that has no specific shape and changes over time, emphasising the variable boundaries (sub-boundaries) of the project phases due to their continuous redefinition as the project progresses. The ability of these phase boundaries to expand, for *sweeping-in* relevant information, and to contract, to avoid its *over-inclusion*, makes *phase-stakeholder-identification* a rigorous ethical exercise. Moreover, like specific SDLC phases (investigation, analysis, design, development and maintenance), this process, based on learning, can also be repeated at any time, as required. This, in essence, does not damage the basic setting of the SDLC as a development methodology since project activities may be repeated at any time for modification and improvement of the system being developed (O'Brien and

Marakas, 2005). *Sweep-in* thus becomes an integral part of the traditional SDLC, making it compulsory to define an ethically justified systemic stakeholder network before initiating a new IS adoption phase.

Thus, the process of phase-stakeholder-identification will help project managers to

- justify/redefine the *roles* (Table 5.1) and *involvement* (active or passive) of each stakeholder before a project phase is initiated (Figure 5.6)
- manage stakeholders by looking deeply into the interactions occurring inside the *systemic stakeholder networks* (Figure 5.5) employing social network mechanisms (Singh, 2005) and/or altering the network structures when required (Cummings and Worley, 1993)
- glide through various phases in, ideally speaking, a conflict-free environment or, at least, one that has a minimum of conflict
- address the problems of systems delivery and communication pertaining to the SDLC (Berrisford and Wetherbe, 1979; Gremillion and Pyburn, 1983) through clear definition of roles and responsibilities and communication management of stakeholder networks.

Limitations and Conclusion

Based on the preceding analysis and discussion, a practical procedure, incorporating critical systems thinking, network stakeholder theory and innovation diffusion, for systematically conducting the IS adoption project has been presented. The model, however, is not a wizard's wand for leading project managers to triumph. Rather, the proposed model is capable of assisting project managers along the road to success through addressing key factors that are essential for a successful IS adoption project.

Traditional approaches (like the SDLC) to IS development overemphasise the design and construction of computer-based artefacts without giving sufficient attention to the social and contextual sides of the process (Avison et al., 1998). A critical systems thinking perspective is beneficial for engaging these in IS adoption in an organisational context. Boundary critique and systemic stakeholder networks provide the knowledge base and the strategic view for managing stakeholder-related issues in IS adoption and their impact on organisations during innovation diffusion. Our model, however, still needs to be empirically tested. Its likely practicality, though, lies in the fact that CST, the SDLC and network mechanisms are procedures or methods that have previously been applied and tested in real-life scenarios.

The model suggests the use of interventions to influence attitudes towards IS adoption to mitigate resistance. But it recommends neither any particular intervention plan nor any measure of effectiveness for an intervention strategy. It also does not provide a mechanism to indicate when these interventions transform from facilitating innovation diffusion to its manipulation.

The model uses the SDLC as a roadmap for an IS adoption project but we do not discuss how the proposed model can be modified to accommodate the phases associated with some other system development methodologies like rapid application development (RAD), joint application development (JAD), the spiral model, and so on. We also do not discuss its applicability or coupling with other technology acceptance models like TAM or TAM2; however, these two aspects are future research directions that can be taken in the development of such models.

References

Achterkamp, M. C., & Vos, J. F. J. (2007). Critically identifying stakeholders. Evaluating boundary critique as a vehicle for stakeholder identification. *Systems Research and Behavioral Science, 24*(1), 3–14.

Ackoff, R. L. (1995). Whole-ing the parts and righting the wrongs. *Systems Research and Behavioral Science, 12*(1), 43–6.

Avison, D. E., Wood-Harper, A. T., Vidgen, R. T., & Wood, J. R. G. (1998). A further exploration into information systems development: the evolution of Multiview2. *Information Technology and People, 11*(2), 124–39.

Azzara, C., & Garone, S. (2003). *IT adoption profiles and application implementation failures*. Research report. AlignIT Group.

Berrisford, T. R., & Wetherbe, J. C. (1979). Heuristic development: a redesign of systems design. *MIS Quarterly, 3*(1), 11–19.

Burt, R. S. (1980). Models of network structure. *Annual Review of Sociology, 6*, 79–141.

Cao, G., Clarke, S., & Lehaney, B. (2003). Diversity management in organisational change: towards a systemic framework. *Systems Research and Behavioral Science, 20*(3), 231–42.

Carugati, A. (2008). Information system development activities and inquiring systems: an integrating framework. *European Journal of Information Systems, 17*(2), 143–55.

Chen, J., & Latendresse, P. (2003). The information age and why IT projects must not fail. Paper presented at the Thirty-Fourth Southwest Decision Sciences Institute Conference, Houston, Tex., March.

Churchman, C. W. (1970). Operations research as a profession. *Management Science, 17*(2), B37–53.

Churchman, C. W. (1979). *The Systems Approach and Its Enemies*. New York, NY: Basic Books.

Cooper, R., & Markus, M. L. (1995). Human reengineering. *Sloan Management Review, 36*(4), 39–50.

Cummings, T. G., & Worley C. G. (1993). *Organization Development and Change*. Minneapolis, Minn.: West Publishing.

Davis, F. D. (1989). Perceived usefulness, perceived ease of use, and user acceptance of information technology. *MIS Quarterly, 13*(3), 319–40.

Davis, F. D., Bagozzi, R. P., & Warshaw, P. R. (1989). User acceptance of computer technology: a comparison of two theoretical models. *Management Science, 35*(8), 982–1002.

Dow, G. (1988). Configurational and coactivational views of organizational structure. *Academy of Management Review, 13*(1), 53–64.

Erickson, B. H. (1988). The relational basis of attitudes. In B. Wellman & S. D. Berkowitz (eds), *Social Structures: A network approach* (pp. 99–121). Cambridge, UK: Cambridge University Press.

Festinger, L., Schacter, S., & Back, K. (1950). *Social Pressures in Informal Groups: A study of human factors in housing*. Palo Alto, Calif.: Stanford University Press.

Freeman, R. E. (1984). *Strategic Management: A stakeholder approach*. Boston, Mass.: Pitman.

Freeman, R. E., & Evan, W. M. (1990). Corporate governance: a stakeholder interpretation. *The Journal of Behavioral Economics, 19*(4), 337–59.

Gremillion, L. L., & Pyburn, P. (1983). Breaking the systems development bottleneck. *Harvard Business Review, 61*(1), 130–7.

Hackman, J. R. (1983). Group influences on individuals. In M. D. Dunnette (ed.), *Handbook of Industrial and Organizational Psychology* (pp. 1455–525). New York, NY: Wiley.

Hall, G. E., & Hord, S. M. (1987). *Change in Schools: Facilitating the process.* Albany, NY: SUNY Press.

Hall, G. E., & Hord, S. M. (2006). *Implementing Change: Practices, principles and potholes.* Boston, Mass.: Pearson.

Hartman, R. L., & Johnson, J. D. (1989). Social contagion and multiplexity: communication networks as predictors of commitment and role ambiguity. *Human Communication Research, 15*(4), 523–48.

Jick, T. D. (1993). *Managing Change: Cases and concepts.* Homewood, Ill.: Irwin.

Kettinger, W. J., Teng, J. T. C., & Guha, S. (1997). Business process change: a study of methodologies, techniques, and tools. *MIS Quarterly, 21*(1), 55–81.

Kroenke, D. M. (2009). *Using MIS.* Upper Saddle River, NJ: Prentice-Hall.

McIlduff, E., & Coghlan, D. (2000). Understanding and contending with passive-aggressive behaviour in teams and organizations. *Journal of Managerial Psychology, 15*(7), 716–32.

Mahmood, M. A. (1987). System development methods—a comparative investigation. *MIS Quarterly, 27*(3), 293–311.

Malmsjö, A., & Övelius, E. (2003). Factors that induce change in information systems. *Systems Research and Behavioral Science, 20*(3), 243–53.

Midgley, G. (2003). Science as systemic intervention: some implications of systems thinking and complexity for the philosophy of science. *Systemic Practice and Action Research, 16*(2), 77–97.

Midgley, G. (2007). Systems thinking for evaluation. In B. Williams & I. Imam (eds), *Systems Concepts in Evaluation: An expert anthology* (pp. 11–34). Point Reyes, Calif.: Edge Press.

Nilakanta, S., & Scamell, R. W. (1990). The effect of information sources and communication channels on the diffusion of innovation in a database development environment. *Management Science, 36*(1), 24–40.

O'Brien, J. A., & Marakas, G. (2005). *Introduction to Information Systems.* New York, NY: McGraw-Hill.

Paper, D., Rodger, J., & Pendharkar, P. (2000). Development and testing of a theoretical model of transformation. *Proceedings of the 33rd Hawaii International Conference on Systems Sciences,* 1–9.

Rogers, E. M. (1995). *Diffusion of Innovations.* [Fourth edn]. New York, NY: The Free Press.

Rogers, E. M., & Kincaid L. D. (1981). *Communication Networks: Toward a new paradigm for research*. New York, NY: The Free Press.

Rowley, T. J. (1997). Moving beyond dyadic ties: a network theory of stakeholder influences. *The Academy of Management Review, 22*(4), 887–910.

Singh, J. (2005). Collaborative networks as determinants of knowledge diffusion patterns. *Management Science, 51*(5), 756–70.

Standing, C., Guilfoyle, A., Lin, C., & Love, P. E. D. (2006). The attribution of success and failure in IT projects. *Industrial Management and Data Systems, 106*(8), 1148–65.

Ulrich, W. (1983). *Critical Heuristics of Social Planning: A new approach to practical philosophy*. New York, NY: Wiley.

Venkatesh, V., & Davis, F. D. (2000). A theoretical extension of the technology acceptance model: four longitudinal field studies. *Management Science, 46*(2), 186–204.

Venkatesh, V., Morris, M. G., Davis, G. B., & Davis, F. D. (2003). User acceptance of information technology: toward a unified view. *MIS Quarterly, 27*(3), 425–78.

Vos, J. F. J. (2003). Corporate social responsibility and the identification of stakeholders. *Corporate Social Responsibility and Environmental Management, 10*(3), 141–52.

Waldman, J. N. (2007). Thinking systems need systems thinking. *Systems Research and Behavioral Science, 24*(3), 271–84.

Wilby, J. (2005). Combining a systems framework with epidemiology in the study of emerging infectious disease. *Systems Research and Behavioral Science, 22*(2), 385–98.

Yardley, D. (2002). *Successful IT Project Delivery*. London, UK: Pearson.

Zaltman, G., Duncan, R. B., & Holbek, J. (1973). *Innovations and Organizations*. New York, NY: Wiley.

6. Advancing Task-Technology Fit Theory: A formative measurement approach to determining task-channel fit for electronic banking channels

Hartmut Hoehle
Victoria University of Wellington

Sid Huff
Victoria University of Wellington

Abstract

Since the advent of contingency theory in organisational research, the notion of 'fit' has continuously grown in importance. The fit concept is evident in a variety of theories in information systems (IS) research as well. In particular, task-technology fit (TTF) theory is recognised as an important development in IS theory. The incorporation of fit constructs into IS models has led to a need to develop reliable and valid methods for measuring fit. In this chapter, we extend TTF theory by proposing and developing a model and measurement approach for task-*channel* fit (a variation on the TTF concept, concerned with electronic banking channels). We thoroughly test our conceptualisation of fit using a series of focus group discussions. Following a parallel-instruments approach, we develop and test a survey instrument for assessing task-channel fit, wherein TCF is modelled as a formative measure. Data gathered from 280 respondents are used to rigorously test the measurement model. Analysis of the data supports the overall soundness of the proposed deviation-score measurement approach. Opportunities for applying this approach in future research are discussed.

Introduction

More than three decades ago, Peter Keen emphasised the need for a 'cumulative tradition' in information systems (IS) research. He suggested that a cumulative tradition requires IS researchers to 'build on each other's work' (Keen, 1980).

Since his call, much attention has been paid to theory development in IS research. Evermann and Tate (2009) noted that prominent instances of successful theory development include the DeLone and McLean (1992) information systems success model, the technology acceptance model (TAM) (Davis, 1989), as well as the unified theory of acceptance and use of technology (UTAUT) (Venkatesh et al., 2003).

Task-technology fit (TTF) theory is also seen as an important development in IS theory (Evermann and Tate, 2009; Goodhue and Thompson, 1995). TTF can be defined as 'the degree to which a technology assists an individual in performing his or her portfolio of tasks' (Goodhue and Thompson, 1995).

Since Goodhue and Thompson's seminal TTF article was published in *MIS Quarterly* in 1995, much research has investigated *fit conceptualisation and measurement* in research disciplines including organisational management (Edwards, 2001), marketing/IS (Jiang et al., 2002) and IS (Chan et al., 1997). Two approaches in particular—deviation-score analysis and parallel instruments—have been frequently used to quantify *fit* between two or more variables (Edwards, 2001; Klein et al., 2009). Deviation-score analysis has been described as a superior fit-assessment technique because 'fit is specified without reference to a criterion variable' (Venkatraman, 1989, p. 430).

Despite a substantial body of knowledge on task-technology fit, to date there have been no rigorous studies assessing TTF via parallel instruments using 'matching' approaches. In the spirit of *building on each other's work*, we propose using a parallel instrument in combination with deviation-score analysis to assess task-technology fit.

We first revisit Goodhue and Thompson's (1995) TTF theory and discuss fit theory and measurement. TTF theory is then applied to electronic banking channels such as ATMs, telephone banking, Internet banking and mobile banking. We term this variant of TTF *task-channel fit (TCF)*. Then we discuss the TCF conceptualisation and explain the instrument development. Afterwards we describe the research methodology and outline how we collected data. The findings are discussed and the chapter concludes by providing managerial and research implications of this study. Finally, directions for future research are suggested.

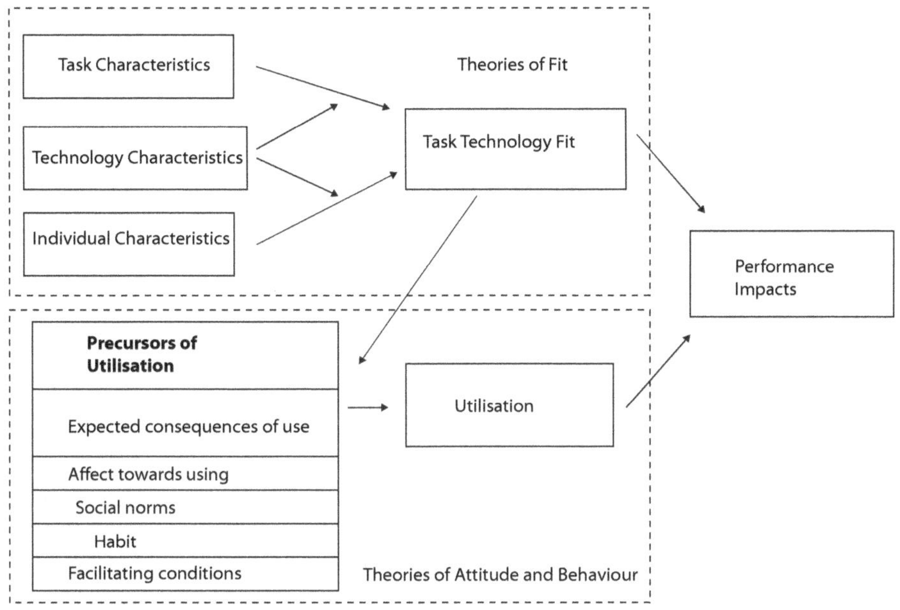

Figure 6.1 Task-to-Performance Chain

Source: Goodhue and Thompson (1995).

Literature Review

Goodhue and Thompson's Task-to-Performance Chain

In order to investigate the linkage between information technology (IT) and user performance, Goodhue and Thompson (1995) conceptualised a *task-to-performance chain* (TPC). This theoretical framework was based on two separate research streams: 1) the user adoption and acceptance research investigating user beliefs and attitudes to predict the utilisation of information systems (Bagozzi, 1982; Baroudi et al., 1986; Davis et al., 1989; Fishbein and Ajzen, 1975; Robey, 1979; Swanson, 1987); and 2) the 'fit focus' evident in research investigating the impact of data representation on the performance of individual IT users (Benbasat et al., 1986; Cooper and Zmud, 1990; Dickson et al., 1986; Jarvenpaa, 1989; Tornatzky and Klein, 1982; Vessey, 1991). Central to this framework was the task-technology fit construct (Goodhue and Thompson, 1995). Figure 6.1 displays the task-to-performance chain framework.

Starting from the left-hand side, the model theorised that task-technology fit is the correspondence between task requirements and individual abilities, moderated by the functionality of the technology (Goodhue and Thompson, 1995). Task-technology fit was predicted to influence the 'precursors of utilisation' and also impact on the performance of the technology user. The conceptualised precursors of utilisation (including expected consequences of use, affect towards using, social norms, habit and facilitating conditions) in turn impacted on technology utilisation, which in turn affected user performance (Goodhue and Thompson, 1995).

The following section discusses literature related to task-to-performance chain/ task-technology fit.

Literature Researching TPC/TTF

Since its introduction, the TPC framework (or parts of it) and, in particular, the TTF construct have been used to study a diverse range of information systems in various contexts (D'Ambra and Wilson, 2004a, 2004b; Dishaw and Strong, 1999; Goodhue, 1998; Goodhue and Thompson, 1995; Karimi et al., 2004; Zigurs and Buckland, 1998; Zigurs et al., 1999).

Goodhue and Thompson (1995) tested a simplified TPC model investigating how TTF influences users in an organisational context. They found support for the hypothesised linkage between the TTF construct and users' performance but not for the causal relationship between the TTF variable and utilisation.

Ferratt and Vlahos (1998) investigated the fit between computer-based information systems (CBIS) and the needs of managers in their decision-making tasks. To measure the TTF, user evaluations of computer-based information systems were used to assess how these systems would support managers in their decision-making process.

Dishaw and Strong (1999) combined the TAM model with the TTF model and tested the extended version in an organisational use setting. The analyses showed that the extended model explained more variance than either TAM or TTF alone (Dishaw and Strong, 1999).

Klopping and McKinney (2004) treated consumer e-commerce as a technology-adoption process and evaluated the suitability of both TAM and TTF to understanding how and why people participate in electronic commerce. To better understand online shopping activity, this study tested a modified TAM model through a web-based survey of 263 undergraduate students (Klopping and McKinney, 2004). The results confirmed that a TTF construct was a valuable addition to the TAM model because the extended model explained more variance in the dependent variable.

Staples and Seddon (2004) tested the technology-to-performance chain in voluntary and mandatory use settings. The entire TPC research model (Goodhue and Thompson, 1995) was tested by surveying university staff (mandatory use) and students (voluntary use) regarding their usage of library services. In both settings, strong support was found for the impact of TCF on performance, as well as on attitudes and beliefs about use (Staples and Seddon, 2004).

While there are numerous other studies based on the TPC framework and the TTF construct, the above-mentioned examples illustrate their wide acceptance within the IS research discipline. Not surprisingly, TTF has even been referred to as 'one of the few prominent theories in our research discipline' (Evermann and Tate, 2009).

Our review of the literature on TPC/TTF indicates that task-technology fit measurement has been operationalised in a variety of different ways. The following section reviews 'fit theory' and discusses how the concept of 'fit' can be measured.

Fit Theory and Measurement

In a seminal article on fit assessment in strategy research, Venkatraman (1989) discussed six alternative measurement approaches for the concept of fit. This section discusses the approaches that are relevant to this study and briefly comments on literature using these techniques.

Fit as Moderation

According to the moderation perspective, the fit between the predictor and the moderator variable is the primary determinant of the criterion variable (Venkatraman, 1989). Figure 6.2 illustrates this conceptualisation of fit.

Mathematically, this can be represented as Equation 6.1.

Equation 6.1

$Z = f(X^*Y)$

In Equation 6.1, Z is the criterion variable, X is the moderator variable and Y is the predictor variable. Researchers applying this approach are assuming that the underlying theory 'specifies that the impact of the predictors (e.g. strategy) varies across the different levels of the moderator (e.g. environments)' (Venkatraman, 1989, p. 424). Moderation can be calculated with regression techniques or ANOVA interaction terms.

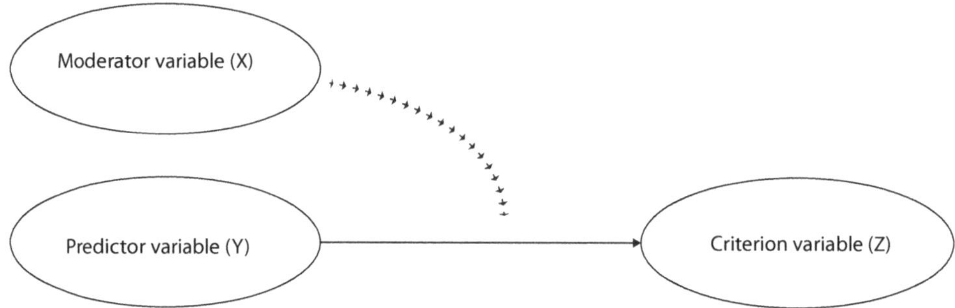

Figure 6.2 Fit as Moderation

Source: Venkatraman (1989).

Chan et al. (1997) applied the moderation approach to investigate IS strategic alignment between business strategic orientation and information systems strategic orientation. They developed a parallel instrument to assess the strategic orientation of business enterprises (STROBE) and the strategic orientation of the existing portfolio of IS applications (STROEPIS) (Chan et al., 1997). Both instruments tapped into eight distinctive strategic dimensions (aggressiveness, analysis, internal defensiveness, external defensiveness, futurity, proactiveness, risk aversion and innovativeness) and for each STROBE item a parallel STROEPIS measure was created. The following example illustrates two parallel items for a particular aspect of business strategic orientation.

Table 6.1 Parallel Items Used to Determine Strategic Alignment

STROBE	We are almost always searching for new business opportunities.
STROEPIS	The systems used in this business unit assist in the identification of new business opportunities.

Source: Chan et al. (1997).

Both items were measured using Likert scales with anchors 5 (strongly agree) to 1 (strongly disagree).

It was assumed that: 'STROEPIS moderated the relationship between STROBE and business performance. In a similar fashion, STROBE moderated the relationship between STROEPIS and IS effectiveness. It was the combination of, or synergy between, STROBE and STROEPIS rather than the difference between the two, that was most important' (Chan et al, 1997, p. 144). In order to calculate the moderation scores, STROBE*STROEPIS product terms were computed (Chan et al., 1997). The STROBE*STROEPIS fit scores were used to assess the structural aspects of the overall research model.

Moderation approaches have also been used to assess task-technology fit. For example, Goodhue (1995) investigated user evaluations of IS via task-technology fit. As part of the TTF model development, the author argued that 'the strength of the link between a system characteristic and user evaluations of it will depend upon how important that characteristic is, given the task demands and the capabilities of the user. This corresponds exactly to one of Venkatraman's categories of fit, fit as moderation' (Goodhue, 1995, p. 1834).

Similarly, Dishaw and Strong (1999) computed TTF by matching task characteristics and the supporting functionality of technology using the moderation (or interaction) approach. While not explained in detail, they argue that '[f]or the TTF model, task-technology fit is computed by matching characteristics of a maintenance task to supporting functionality in a software maintenance tool, using an interaction approach' (Dishaw and Strong, 1999).

Fit as Mediation

The fit as mediation perspective assumes the 'existence of a significant intervening mechanism (e.g. organisational structure) between an antecedent variable (e.g. strategy) and the consequent variable (e.g. performance)' (Venkatraman, 1989, p. 428). Fit as mediation can be represented visually as shown in Figure 6.3.

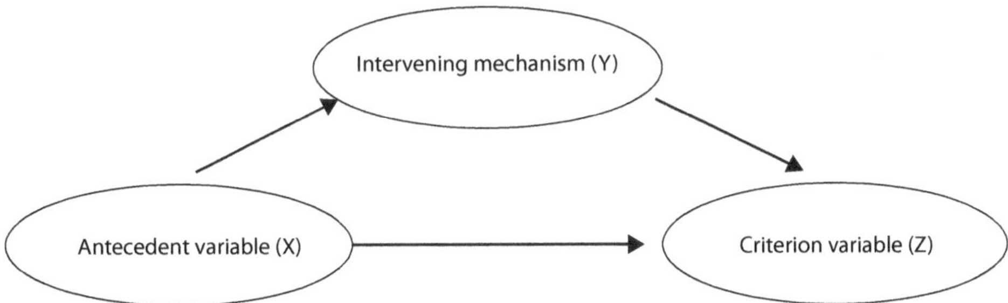

Figure 6.3 Fit as Mediation

Source: Venkatraman (1989).

As with moderation, this perspective is anchored with respect to a specific criterion variable; however, fit is viewed as indirect, making it less precise than the moderation perspective. The mediator variable (Y) accounts for a significant proportion of the relation between the predictor (X) and criterion variable (Z) (Venkatraman, 1989).

Thatcher (2001) studied the extent to which communication media and demographic diversity predict creativity. Identity-fit was predicted to mediate the relationships between the antecedents and creativity (Thatcher, 2001). The authors used hierarchical regression analyses to measure the mediation effects.

It should be noted that strategy researchers have traditionally embraced the moderation approach rather than using the mediation approach to assess fit (Venkatraman, 1989).

Fit as Matching

This perspective of fit suggests that fit is a theoretically defined match between two related variables (Venkatraman, 1989). Fit as matching is illustrated in Figure 6.4.

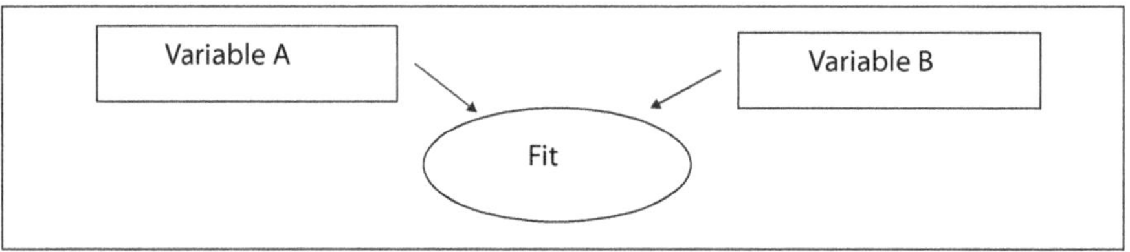

Figure 6.4 Fit as Matching

This approach is 'a major point of departure from fit as moderation and fit as mediation because fit is specified without reference to a criterion variable,' (Venkatraman, 1989, p. 430). Fit as matching can be operationalised using deviation-score analysis or regression residuals. The deviation-score analysis is based on the assumption that 'the absolute difference between the standardised scores of two variables indicates a lack of fit' (Venkatraman, 1989, p. 431).

This form of fit assessment has frequently been applied in organisational, marketing and IS research. Most commonly, this approach is operationalised via parallel instruments. Parallel instruments can be used to collect responses for variable A (see Figure 6.4) separately from variable B. By comparing the responses obtained for variable A and variable B, a fit score can be computed (Edwards, 2001; Klein et al., 2009).

For example, Jiang et al. (2002) applied the ServQual instrument to study service quality in an IS setting. The original ServQual instrument consisted of five distinct dimensions (tangibles, reliability, responsiveness, assurance and empathy) and can be defined as the gap (or fit) between consumer expectations and perceived delivery (Jiang et al., 2002; Zeithaml et al., 1990). To assess this gap, Jiang et al. (2002) obtained a sample of IS professionals and matched IS users. Each respondent group answered parallel questions regarding their service expectations and actual service quality perceptions. Using deviation-score analysis, the authors computed the fit between service expectations and actual experience. Service quality was measured 'by the gap score (G), where G is the difference between corresponding perception of delivered service (P) and expectation of service (E) for each item (G=P–E)' (Jiang et al., 2002, p. 146).

Fit as Gestalts

Venkatraman (1989) suggested conceptualising 'fit as gestalts' when more than two variables are used. Gestalts could be defined as 'the degree of internal coherence among a set of theoretical attributes' (Venkatraman, 1989, p. 432). Gestalts could be arrived at by cluster analysing data (Venkatraman, 1989). Only a few studies in the IS research literature were identified that applied the fit-as-Gestalts approach (Buttermann et al., 2008; Lefebvre et al., 1997). Since this form of fit does not apply to the current study (TTF/TCF does not involve a 'set of theoretical attributes'), it is not further discussed here.

Fit as Profile Deviation

From the profile deviation perspective, fit is 'the degree of adherence to an externally specified profile' (Venkatraman, 1989, p. 433). This perspective of fit differs from the Gestalt perspective in that the 'profile' is anchored to a specific criterion, such as performance (Venkatraman, 1989). Evaluating fit as profile deviation is particularly useful for testing the effects of environment–strategy co-alignment since multiple variables are involved. Using interaction terms or moderating effects of variables can become cumbersome and problematic when multiple variables are involved (Sabberwal and Chan, 2001). Fit as profile deviation can be operationalised using pattern analysis, as demonstrated in a business alignment study by Sabherwal and Chan (2001). Since fit as profile deviation does not apply to the current study, it is not further discussed here.

Fit as Co-Variation

When fit is conceptualised as co-variation, 'fit is a pattern of covariation or internal consistency among a set of underlying theoretically related variables' (Venkatraman, 1989, p. 435). Co-variation can be computed using either exploratory or confirmatory factor analysis. Fit as co-variation involves identifying several dimensions based on the scores along a set of chosen variables.

This form of fit assessment has frequently been used by researchers investigating IS, including studies of task-technology fit. For example, Goodhue and Thompson (1995) assembled 48 items representing aspects of the fit between the tasks users perform and the technologies they use to perform these tasks. Using exploratory factor analysis, the authors first excluded 14 items and collapsed the remaining TTF measures into eight unique factors (quality, locatability, authorisation, compatibility, product timeliness, ease of use, ease of training and relationship with users). They argued that each dimension would represent a unique part of the task-technology fit. Using regression techniques, these facets of fit were linked to other constructs within the research model (for example, utilisation and performance impacts).

Staples and Seddon (2004) also used a multifaceted measure to identify a TTF within the context of their study. They used four facets of TTF originally proposed by Goodhue and Thompson (work compatibility, ease of use, ease of learning and information quality). To test these dimensions, 12 questions (three questions per facet) were used. The authors modelled the TTF construct as a second-order factor, with each facet of TTF being a first-order factor that formed the second-order factor.

D'Ambra and Wilson (2004a) also used a multidimensional construct to model the fit between Web usage and personal travel planning and purchase of flight tickets. To evaluate the TTF construct, the authors developed multiple items that specified the TTF in the context of the study. Next, the authors collected data and factor analysed the TTF items. Several TTF dimensions (uncertainty reduction, fun/flow, mediation, control, information resounds and locatability of information) were identified and used as first-order constructs. Next, the authors used partial least squares (PLS) to assess the structural relationships between these first-order dimensions and utilisation/performance impacts (D'Ambra and Wilson, 2004a).

While not specifically discussed by Venkatraman (1989), 'direct' measures to assess fit have also been used by many researchers. This approach is discussed below.

Direct (Reflective) Fit Measures

The direct measurement approach involves developing and utilising several reflective items that are tailored to elicit individuals' perceptions of the fit between two (or more) variables.

For instance, Bhattacherjee (2001) asked respondents to match their prior expectations to perceived performance of a system. The expectation/confirmation construct was assessed through the following items: '1) My experience with using [the system] was better than what I expected, 2) The service level provided by [the system] was better than what I expected, and 3) Overall, most of my expectations from using [the system] were confirmed' (Bhattacherjee 2001, p. 370).

Similarly, researchers studying TTF have used direct measurement approaches. For example, Klopping and McKinney (2004) created eight reflective items to assess the fit between Internet-based shopping malls and individuals' shopping preferences. The TTF construct was assessed through the following items.

1. Sufficiently detailed product information is maintained on product web sites.

2. On the web sites I visit, product information is either obvious or easy to find out.
3. I can get product information quickly and easily from a web site when I need it.
4. The online product information that I use or would like to use is accurate enough for my purposes.
5. The online product information is up to date for my purposes.
6. The online product information that I need is displayed in a readable and understandable form.
7. The online product information maintained at web sites is pretty much what I need to carry out my tasks.
8. The product information is stored in so many forms it is hard to know how to use it effectively (Klopping and McKinney, 2004).

They used the eight items to construct a scale to measure TTF and then applied structural equation modelling to test a research model that hypothesised relationships between the TTF construct and other variables such as perceived usefulness and intention to use (Klopping and McKinney, 2004).

Similarly, Ferratt and Vlahos (1998) used five direct TTF items to evaluate how computer-based information systems (CBIS) fit to support managers in their decision-making tasks. The measures were designed to assess managers regarding their decision-making habits.

The main advantage of the direct measurement approach is its simplicity. Including a set of reflective measures within a survey questionnaire instrument is straightforward. These constructs can be treated as reflectively measured latent variables, and component (PLS) or co-variance-based (Amos, Lisrel, EQS, and so on) structural equation modelling techniques can be used to evaluate the research models.

Despite the widespread acceptance of this approach, this technique has been criticised by various researchers (Edwards, 2001; Klein et al., 2009). Asking respondents about the perceived direct fit between two or more variables requires the respondents to conceive and mentally 'calculate' their perceptions of fit. Researchers must rely on the respondents' ability to reliably conduct this mental arithmetic as they respond to the fit questions (Kristof, 1996).

Literature Review: Summary and identification of research gap

Table 6.2 summarises the various approaches researchers have employed to assess fit in different research settings.

Table 6.2 Fit Conceptualisation Used in Reference Disciplines and IS Research

	Reference disciplines/IS research	TPC/TTF
Fit as moderation	(Chan et al., 1997; Parker and van Witteloostuijn, 2010; Prescott, 1986)	(Dishaw and Strong, 1999; Goodhue, 1995)
Fit as mediation	(Parker and van Witteloostuijn, 2010; Thatcher, 2001)	n.a.
Fit as matching	(Jiang et al. 2002; Tesch et al., 2003)	n.a.
Fit as Gestalts	(Buttermann et al., 2008; Lefebvre et al., 1997)	n.a.
Fit as profile deviation	(Conrad et al., 1997; Parker et al., 2010; Sabberwal and Chan, 2001)	n.a.
Fit as co-variation	(McKinney et al., 2002; Mitchell et al., 2007)	(D'Ambra et al., 2004a; Goodhue and Thompson, 1995; Staples and Seddon, 2004)
Fit directly assessed	(Bhattacherjee, 2001; Limayem et al., 2007; Lin et al., 2005)	(Ferratt and Vlahos, 1998; Klopping and McKinney, 2004)

Each of the fit conceptualisations should be carefully scrutinised before applying it in a specific research context. For instance, *fit as mediation* assumes that the fit variable has a mediating effect on the dependent variable. Given Goodhue and Thompson's (1995) TPC model, this fit conceptualisation seems to be inappropriate for studying TTF due to the underlying theoretical assumptions. Goodhue and Thompson (1995) theorised that TTF would influence certain precursors of utilisation (such as attitude towards technology). Reversing this causal relationship would be theoretically unjustified and illogical (for example, stating that TTF would be influenced by users' attitudes).

Likewise, fit as Gestalts would be inappropriate for measuring TTF since only three variables 'Gestalt' (*Gestalt* is the German word for 'forming') the fit between a given technology, specific tasks and individuals' attributes. The same underlying principle applies to fit as a profile deviation (as explained previously). Thus, *fit as Gestalt* as well as *fit as profile deviation* should be carefully scrutinised before applying these approaches in investigations of TTF theory in IS contexts.

Further, despite the fact that *fit as co-variation* has been often used in IS research, there are conceptual issues with this approach. For instance, Staples and Seddon (2004) used this technique to assess *task performance chain* theory. Building upon Goodhue and Thompson's (1995) work, Staples and Seddon (2004) conceptualised four different TTF dimensions (work compatibility, ease of use, ease of learning and information quality). Each TTF dimension was measured through three reflective items. The dimensions were then used as first-order constructs comprising the second-order TTF variable. Next, statistical correlations were investigated between the TTF construct and the remaining variables in the research model. This approach has conceptual flaws as the fit is not specified numerically when investigating causal relationships with other variables. For instance, the authors collected data for three *ease of use* items as part of their TTF conceptualisation: '1) the system is easy to use, 2) the system is user friendly, 3) it is easy to get the system to do what I want it to do' (Staples and Seddon, 2004, p. 34). When scrutinising these items carefully, it becomes clear that *tasks* users perform were not considered in these items. The responses collected for these items might, however, co-vary with other variables in the research model (and perhaps the dependent variable–system utilisation), and it appears to be problematic to derive TTF from non–task-specific *ease of use* items. It is important to note that multiple other TTF-related studies (for example, Goodhue and Thompson, 1995) also used *ease of use* items to specify TTF.

Directly assessing fit can also be problematic (Venkatesh and Goyal, 2010) because 'measures that elicit direct comparisons merely shift the onus of creating a difference score from the researcher to the respondent' (Edwards, 2001, p. 268). Researchers commonly use response scales (ranging from negative to positive numbers) to collect data and they ask the respondent to mentally calculate the difference of the fit components themselves.

Finally, fit researchers have also argued that *fit moderation* approaches should be treated with care. Edwards (2001) expressed concern that many researchers have resorted to product terms when confronted with problems with difference scores: 'The use of product terms as a substitute for difference scores is alluring, given that product terms analysed hierarchically capture the interaction between two variables, and the terms interaction and fit often have been used jointly, if not interchangeably, in congruence research.' While not discussed in detail here, Edwards (2001, p. 270) provides an in-depth discussion of why there are mathematical issues when replacing deviation scores through product terms.

Given these arguments, using deviation-score analysis in combination with a parallel instrument appears to be a superior technique for measuring TTF in an IS research context. Using a parallel instrument would allow collecting responses for 'both sides' of the fit construct separately. The deviation-score analysis could be used to match the responses without 'priming the respondent to mentally subtract the components' (Edwards, 2001, p. 268).

Despite the relatively large number of research studies on task-technology fit, to our knowledge, no study has attempted to quantify the fit between a given technology and a specific task via deviation-score analysis (*matching* approach). Thus, we report here a study of task-channel fit (a variation on TTF) in which we develop a deviation-score approach to measuring fit, utilising parallel instruments. The task-channel fit conceptualisation is now explained in more depth.

Task-Channel Fit Conceptualisation

Since the early 1970s, the proliferation of new information and communication technologies within the financial industry has significantly influenced the way banks service consumers. In particular, self-service technologies have enabled banks to pursue an electronically mediated multi-channel strategy. Nowadays ATMs, telephone banking, Internet banking and mobile banking are all efficient means for selling products and servicing customers.

For the consumer, these electronic banking channels largely eliminate the need to visit a branch and offer convenient access to bank accounts. Banks also benefit from self-service technologies as they can cut costs incurred by the traditional branch network.

Usage rates suggest, however, that banks are missing out on the opportunity to move even more customers to electronic banking services. For example, each month 73 per cent of all European banking customers use ATM machines, although only 24 per cent use Internet banking services (Deutsche Bank Research, 2006). Similarly, although most North American and Australasian retail banks offer phone banking and mobile banking services, only 5–10 per cent of all consumers have used them (Forrester Research, 2007).

Moreover, consumers favour specific electronic banking technologies for specific product categories. For instance, Internet banking applications are used for simple product categories (for example, domestic transactions) as well as more complicated product categories such as international payments, credit card applications and financial loans (Deutsche Bank Research, 2006; Forrester Research, 2007). In contrast, complex financial transactions are seen to be difficult to perform on mobile phones due to their hardware limitations such as small screens and clumsy input mechanisms. In consequence, consumers tend to use mobile devices for simple banking transactions in situations where they need instant access to their accounts and other banking channels are not in reach (for example, checking their account balance before purchasing goods at a point of sale).

These varying usage patterns indicate that each electronic banking channel has inherent capabilities that align with certain types of banking tasks—and clash with others. This suggests the notion of a 'fit' between a given electronic banking technology and specific banking tasks. Furthermore, it seems reasonable to assume that the better the fit between electronic banking technology and banking task, the higher will be the adoption and utilisation of the service.

Despite a substantial body of knowledge on electronic banking services, to date there have been no rigorous studies investigating the fit between electronic banking channels and banking tasks. In an attempt to address this gap, we have conceptualised task-technology fit for electronic banking channels (which we refer to as task-channel fit, or TCF) and have operationalised a deviation-score (matching) approach for measuring this construct.

Task-Technology Fit of Electronic Banking Technologies

Drawing from the task-technology fit definition, task-channel fit is defined as *the user's perception of the suitability of a particular electronic banking channel to support a particular banking task*.

Banking tasks include the various kinds of financial and non-financial transactions a consumer wishes to conduct with his or her bank. The existing literature suggests that banking tasks can be characterised along a variety of dimensions. Five such dimensions have been identified.

Dimension One: Task complexity

Several studies using the TTF concept categorised tasks into simple versus complex tasks (Shirani et al., 1999; Zigurs and Buckland, 1998). For example, Zigurs and Buckland (1998) emphasised the importance of task complexity when considering task-technology fit for group support systems. As the literature has shown, banking tasks also vary in their complexity (for example, account inquiries are considered simple tasks, while securing a financial loan is a complex task) (Sayar and Wolfe, 2007; Tan and Thompson, 2000).

Dimension Two: Task effort

Using Wood's (1986) and Campbell's (1988) task complexity frameworks, Nadkarni and Gupta (2007) argued that certain tasks require a considerable amount of effort without requiring much cognitive workload from the person performing the task (Campbell, 1988; Nadkarni and Gupta, 2007; Wood, 1986). Likewise, conducting some financial transactions entails more effort than others without necessarily being more complicated. For instance, filling out forms for international transactions is as conceptually simple as conducting domestic transactions, but a bank usually requires much more information to process overseas remittances.

Dimension Three: Task frequency

Researchers in various disciplines have investigated how recurring behaviour impacts on individuals' actions (Rangan et al., 1992; Reinsch and Beswick, 1990). For instance, Rangan et al. (1992) argued that frequency of purchase impacts on consumers' channel selection. Behavioural frequency has also been noted by IS researchers studying the impact of regular or habitual system usage (Limayem et al., 2007; Ortiz de Guinea and Markus, 2009). In the context of electronic banking, consumers perform certain banking tasks more often than others. For instance, most individuals check their account balances frequently while they seldom apply for mortgages.

Dimension Four: Task importance

Consumers view certain transactions as being more salient than others (Reinsch and Beswick, 1990). For example, high-value transactions (like those involving hundreds of dollars) are often viewed as more important than transactions with very low values (Sayar and Wolfe, 2007). Also, transactions such as mortgages or financial loans impact significantly and over a longer time span on a consumer's personal life, hence are perceived as being of high importance, while account inquiries are often seen as low-importance tasks.

Dimension Five: Task time criticality

The level of time criticality is a fifth important aspect of financial transactions (Kleijnen et al., 2004; Liao and Cheung, 2002; Tan and Thompson, 2000). Financial transactions such as foreign exchange trades or share purchases are highly time sensitive due to market volatility, and often require immediate execution. On the other hand, tasks such as transfers, loan applications or insurance acquisitions are less time critical for consumers.

Task-channel fit, then, is conceptualised as the aggregate correspondence between a consumer's perception of the characteristics of a banking task (in terms of the five dimensions above) and the suitability of a particular banking channel to support a banking task with those characteristics. Figure 6.5 illustrates the TCF concept.

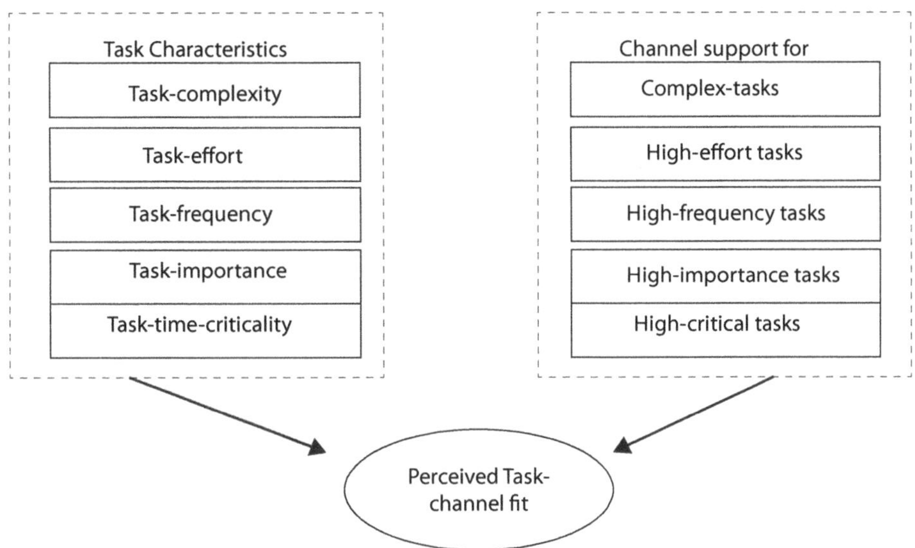

Figure 6.5 Perceived Task-Channel Fit Conceptualisation

Since these dimensions were solely based upon the existing literature, focus group sessions were conducted to further validate the TCF dimensions.

Focus Group Discussions

Five exploratory focus groups (each consisting of five electronic banking users) were carried out. The primary goal of these sessions was to develop an understanding of how consumers perceive the task-channel fit of electronic banking channels. Theoretically motivated purposive sampling methods were employed in selecting participants for this study (Calder, 1977). The focus groups were selected so as to achieve a wide variety of individual characteristics across the different user groups.

Research participants were contacted via email, telephone or face-to-face conversations. A prerequisite for participation within the focus groups was that each participant had used at least one electronic banking channel (ATM, telephone, Internet or mobile banking) for their banking needs and that they were familiar with the most common functionalities of that specific channel. Also, all participants were involved in the purchasing of financial products in their own household. The age of participants ranged between twenty-four and sixty-one years and the focus groups included 11 men and 15 women. Educational levels ranged from high school education to PhDs in engineering science.

Data collection was carried out through focus group discussions featuring open-ended questions. An example of the questions presented to the focus group panels is given below.

> What banking transactions/products do you perform on each electronic banking channel, and why?

The focus group discussions lasted between 60 and 90 minutes each and were recorded and transcribed afterwards. To analyse the data, coding techniques commonly used for grounded theory studies were applied (Strauss and Corbin, 1998). The codes were visualised in a data matrix to highlight similarities and differences between the various codes (Miles and Huberman, 1994).

Overall, the focus group discussions confirmed the task-channel fit dimensions described above. During the focus group discussions, task complexity, task effort, task frequency, task importance and task time criticality were repeatedly identified as important factors influencing the suitability of electronic banking channels to support specific banking tasks.

For instance, many respondents argued that electronic banking channels differ in terms of their ability to support complex banking tasks. One participant stated:

> Telephone banking is much less developed than Internet banking. I'd say it is much more archaic and I used telephone banking back in the days when I was a student. I only had one account and never used it to transfer money between accounts. I use Internet banking for all my transactions nowadays. I could never do on telephone banking what I do on the Internet banking; telephone banking is just not suited for more complicated banking services.

The level of time criticality was considered another important aspect of financial transactions. Most participants indicated that they perceived some banking transactions as urgent while others were seen as less time critical. In addition, some electronic banking channels were seen as supporting time-critical tasks particularly well (for example, Internet and mobile banking) whereas others were seen as less suited for urgent matters (for example, ATM banking).

> The other day I was out at a bar and realised that I hadn't paid my phone bill. That was an urgent matter for me so I just used my text-banking and paid the bill. That's 24 and on the go...so, for me, mobile banking is something for quick and easy day-to-day tasks.

The focus groups also suggested that consumers execute some banking tasks more frequently than others. Depending on the frequency of performing a given banking task, consumers would view certain banking channels as more suited than others. The participants also indicated that they would develop a routine if they performed tasks regularly on a given banking channel, which would influence their perceptions of that channel. After they repeated the banking tasks several times on a specific channel, they would perform these tasks automatically because of learning. Task frequency was also a recurring concept within most focus group discussions.

One participant argued:

> I think regularity and how often you have to perform each banking task is very important when using electronic banking channels. Once you have learned how to use them and if you do them frequently...the process becomes habitual. For example, I know exactly which buttons to press when using ATMs or telephone banking. That's because I use them quite often. The procedure stays the same and I know exactly what to do.

In summary, the focus group discussions supported the initial conceptualisation of the task-channel fit of electronic banking channels. The following section describes the instrument development for measuring the TCF construct.

Instrument Development

This research intended to create two sets of TCF measures—one formative and one reflective. The formative set aimed to measure specific dimensions of the task-channel fit construct. These items intended to capture different aspects of TCF and should individually represent a finer level of measurement granularity compared with the reflective items.

When measuring constructs via formative indicators, it is important to understand that the content validity of a construct correlates highly with the indicators used to measure the construct. Each item contributes to 'form' the construct so it is essential that the entire domain is covered (Diamantopoulos et al., 2008; Petter et al., 2007). To assure that each TCF dimension was captured adequately, two items were included for each TCF dimension (displayed in Table 6.3).

We also included four global TCF items. These items were intended to measure consumers' views on the overall fit between Internet banking and financial loans/account inquiries. Since these items were designed to tap into consumers' overall perception of the TCF, the measures were necessarily somewhat more abstract.

While there are issues with measuring the TCF directly through a set of reflective items (see the discussion in the 'Literature Review' subsection above), nevertheless we decided to include the direct TCF measures as they allowed us to check the reliability of the formative measurement approach. The construct reliability of a formative construct can be evaluated through the use of a two construct model that integrates an additional phantom variable, which represents the construct's reflective operationalisation (Diamantopoulos and Winklhofer, 2001; Goetz et al., 2010).

Goetz et al. (2010) elaborated on this and suggested that product quality can be measured by means of formative indicators such as 'appealing design', 'high-quality functionality' and 'adequate product weight'. Product quality can also be measured through reflective indicators to determine the formative operationalisation's error term. Such items could include 'the product is of high quality', 'my quality expectations have been met', 'I will not complain about the product' and 'my quality expectations have been exceeded' (Goetz et al., 2010).

For the current study, the formative TCF measures intended to specify why consumers perceive a fit between a given banking task and an electronic banking channel. This approach is much more detailed and provides a more robust and rich picture of the TCF construct. In contrast, the set of reflective items provides a higher-level overview of perceived TCF.

We regard the formative measurement approach as superior since it provides a more detailed picture (and also avoids the cognitive load on respondents of having to calculate or otherwise determine TCF). Since this is the first attempt to formatively measure the perceived TCF of electronic banking channels, it seemed reasonable, however, to retain both sets of measures until more is known about the relationship of each TCF dimension and the perceived TCF construct.

The development of the measurement scales was conducted in three stages. We first screened the existing literature for items that had been validated by prior research. No items could be identified that matched the construct dimensions. Therefore, we developed new items for the TCF dimensions based on their construct definitions.

To do this, a spreadsheet was established listing the TCF construct definitions. Next, the existing literature and transcripts of the focus group discussions were reviewed for potential keywords describing the TCF dimensions. For instance, time-critical tasks were often referred to as being 'urgent' in the existing literature (Gattiker and Goodhue, 2005; Jenkins et al., 1971; Junglas et al., 2009; Landry et al., 1991; Park et al., 2008; Yuan et al., 2010). Hence, items tapping time criticality were created to reflect that dimension.

Likewise, many focus group participants suggested that certain banking tasks required instant execution and should be performed immediately. Hence, the concepts of 'instant execution' and 'immediacy' were also integrated into time-criticality items. The same procedure was followed for the remaining TCF dimensions.

The initial pool of TCF items was carefully reassessed for conceptual similarities and the wording of each item. For instance, the initial items included the following measures

- a financial loan is a complicated banking transaction
- a financial loan requires a lot of time.

While the first item clearly tapped into the task complexity dimension, the second item—while originally intended to also measure task complexity—was later recognised as being merely concerned about the length of time individuals require to perform the banking task, which is not necessarily a reflection of complexity. Due to this, it was decided to exclude the latter item. Similar steps were performed for the remaining items/TCF dimensions.

The second stage of the instrument development involved two judgment rounds utilising experts relevant to the study's context. The main goal of these expert rounds was to assess the content validity of the scales as well as the wording of the items. The eight judges included two marketing professors, two senior IS researchers, two bank staff and a finance professor as well as a currency trader. The judgment rounds were organised as face-to-face interviews lasting between 60 and 90 minutes each. Each judge was asked to evaluate the content validity of the TCF dimensions as well as to re-examine the items collected for this study. Subsequent to the interviews, the scales were refined appropriately in light of the experts' recommendations.

The third stage of scale development involved two pre-tests of the survey questionnaire instrument. The first pre-test involved five university staff (two administrative staff, one academic staff member and two PhD students) who were asked to complete the survey questionnaire in paper form. Subsequently each respondent was interviewed and questioned about whether they found items unclear or ambiguous, or if they felt confused by some sections of the questionnaire. This feedback was then used to adjust the survey questionnaire instrument. The second pre-test included 15 university staff/PhD students researching information systems. These participants were asked to test the online survey and provide feedback about the structure of the survey and wording of the items. The second pre-test led to the final measures used for this research, summarised in Table 6.3.

Data Collection

In an ideal situation, the TCF scales would be tested by gathering data from respondents for all electronic banking channels, and for a variety of banking tasks; however, this research design appeared infeasible for two reasons. First, combining items for a number of banking transactions (for example, account inquiries, domestic transactions, international payments, applying for credit cards and/or mortgages) with four electronic banking channels (ATMs, telephone, Internet and mobile banking) would lead to a very repetitive and lengthy survey questionnaire. Second, due to varying adoption rates, it seemed unlikely that respondents would be able to reply to questions related to all electronic banking channels.

Table 6.3 Items Used in Constructing the Construct Measures

Construct	Items
TCF—task characteristics	
COMP1	A financial loan (account inquiry) is a complicated banking transaction.
COMP2	Applying for a financial loan (account inquiry) is an easy-to-do banking task.
EFFORT1	I have to provide a lot of information to my bank when applying for a financial loan (account inquiry).
EFFORT2	A loan application (account inquiry) is a banking transaction that requires filling out many forms.
FREQ1	I often apply for a financial loan (account inquiry).
FREQ2	A loan application (account inquiry) is a banking transaction I frequently do.
IMPORT1	A loan application (account inquiry) is an ordinary banking transaction to me.
IMPORT2	Applying for a loan (account inquiry) is a common banking task.
TIME1	I seldom face situations in which I need to apply for a bank loan (account inquiry) urgently.
TIME2	I often need to apply for a financial loan (account inquiry) immediately.
TCF—channel suitability	
	Internet banking is well suited for:
CCOMP1	Complicated banking transactions.
CCOMP2	Easy-to-do banking tasks.
CEFFORT1	Banking transactions for which I have to provide a lot of information to my bank.
CEFFORT2	Banking transactions that do not require filling out many forms.
CFREQ1	Banking transactions I perform often.
CFREQ2	Banking transactions I perform frequently.
CIMPORT1	Ordinary banking transactions.
CIMPORT2	Common banking tasks.
CTIME1	Urgent banking transactions.
CTIME2	Banking transactions I have to do immediately.
Task-channel fit (direct measures)	
TCF1	I think Internet banking would be well suited for loan applications (account inquiry).
TCF2	Internet banking would be a good medium for loan applications (account inquiry).
TCF3	Internet banking would fit well for loan applications (account inquiry).
TCF4	I think Internet banking would be a good way to apply for financial loans (account inquiry).

For this study, then, we focused on a single banking channel and on two different banking tasks. Internet banking was selected as the banking channel. First, most consumers in New Zealand have experience with Internet banking applications and should have well-formed beliefs about most common functionalities of these services. Second, all New Zealand banks offer a wide range of financial products via Internet banking, including simple, medium and complex banking products.

In order to create a meaningful comparison, account inquiries (checking account balance, viewing transaction history, inspecting account statements, and so on) and financial loan applications (applying for bank overdrafts, home loans, personal loans, mortgages, and so on) were selected to test the perceived TCF scales regarding Internet banking services. We operationalised the data collection by using two different versions of the questionnaire instrument (one for account inquiries and another for financial loan applications). The two versions differed slightly, reflecting the nature of the corresponding banking task. Table 6.3 lists the items used to assess the TCF for financial loan applications. As indicated within the table, 'loan applications' was replaced with 'account inquiries' for the second version of the survey questionnaire.

The survey questionnaire instrument was administered to students, faculty and administrative staff at a university in New Zealand. To encourage participation within the study, the survey questionnaire was posted within a university newsletter that is sent out weekly to all staff and students. We received 315 responses. A number of research participants indicated that they felt unfamiliar with either the banking task in question or Internet banking services. These responses were excluded from the data analysis, resulting in 140 responses for each banking task (280 in total).

Data Analysis

Regression Analysis of the Fit Components

Before discussing the TCF assessment, the two 'sides' of TCF—task characteristics and channel suitability for a particular characteristic—were evaluated using linear regression.[1] Klein et al. (2009) suggested analytical guidelines for assessing difference scores in IS research. As part of this, the authors proposed examining

1 There was no compelling theoretical or empirical reason to expect nonlinearity. Therefore, we assumed a linear relationship between the component scores.

whether both component scores of the fit have dissimilar weights. Unequally weighted fit components would indicate that a deviation-score analysis would lead to more significant results (Klein et al., 2009).

We followed these recommendations and conducted a linear regression to assess whether the two components of the TCF (task characteristics and channel suitability) would be unequally weighted. SPSS (Version 17.0) was used to perform the regression analysis. Table 6.4 shows the results. One of the direct measurement items was used as the dependent variable to perform the regression analysis.

Table 6.4 Linear Regression Coefficients/Separate Fit Component Analysis

	Separate components	
	Task characteristics	**Channel suitability**
Loan/complexity	3.907*** (T: 6.721)	1.398 (T: 1.869)
Loan/effort	3.937*** (T: 9.069)	2.668*** (T: 5.038)
Loan/frequency	3.014*** (T: 13.672)	1.582 * (T: 2.359)
Loan/importance	2.797*** (T: 7.737)	1.315* (T: 1.967)
Loan/time criticality	3.548*** (T: 7.823)	1.947*** (T: 3.519)
Acc. inquiry/complexity	3.202*** (T: 5.203)	2.221*** (T: 3.982)
Acc. inquiry/effort	5.209*** (T: 28.575)	3.756*** (T: 8.280)
Acc. inquiry/frequency	3.040*** (T: 9.662)	1.596** (T: 3.013)
Acc. inquiry/importance	2.786*** (T: 4.562)	1.625** (T: 2.920)
Acc. inquiry/time criticality	4.657*** (T: 9.749)	2.943*** (T: 6.506)

* significant at $p < 0.05$

** significant at $p < 0.01$

*** significant at $p < 0.001$

The results confirm that the two components of the task-channel fit are unequally weighted. For both banking tasks (financial loans and account inquiries), there is an unequal weight distribution between both fit components (task characteristics and channel suitability). All task-characteristic items were highly significant; however, certain of the channel suitability sub-components were less significant (for example, loan/frequency).

TCF Calculation

As indicated earlier, we felt that computing a fit score by matching the responses obtained via the parallel instrument approach would yield a richer and more robust measure of TCF than attempting to assess it directly via a set of reflective indicators (Chan et al., 1997; Venkatraman, 1989). Therefore, the task characteristic and channel suitability items (see Table 6.3) were designed to parallel each other. For each individual task characteristic item, a parallel channel suitability item was created so as to allow us to determine the extent of the fit (or lack thereof) the respondent perceived between the task and the channel for that specific task dimension. For instance, COMP1 asks individuals whether they view financial loan applications as complicated. CCOMP1 then inquires whether the individual views Internet banking as well suited for complicated banking transactions. The responses to these two items can be compared to determine an indicator of the fit of the particular channel to that aspect, or dimension, of the particular task. The larger the difference (either positive or negative) between the two ratings, the lower is the degree of fit.

Following this approach, fit scores were calculated for each TCF dimension (for example, TCFcomp2, TCFeffort1, TCFeffort2, and so on). These values were subsequently used as formative indicators for TCF, as they 'formed' the perceived TCF construct. The details of how this was done are explained next.

Data Analysis

Partial least squares (PLS) was used to analyse the data. Partial least squares simultaneously analyses how well the measures relate to each construct and whether the hypotheses at the theoretical level are true. In contrast, with co-variance-based structural equation modelling (SEM) techniques, PLS can handle formative indicators that are required to evaluate the TCF construct using the matching approach. SmartPLS was selected as the software package to perform the data analysis.

The data analysis for the TCF model was assessed for account inquiries and financial loans separately. Aggregating the answers for both versions of the questionnaire would not be meaningful since the perceived TCF would be expected to differ from task to task.

The matching scores for this study were computed for both datasets and used as formative measures for the perceived TCF construct, as illustrated in Figure 6.6.

As suggested by Cenfetelli and Bassellier (2009), all formative measures were initially assessed for multi-collinearity. In contrast with reflective items, multi-collinearity is undesirable for formative measures. Variance inflation factor (VIF) statistics were used to assess all items. A VIF of three or greater indicates the presence of a significant degree of multi-collinearity among the items (Petter et al., 2007). Variance inflation factor statistics can be computed using linear regression methods using SPSS. Table 6.5 shows the scores obtained for these tests.

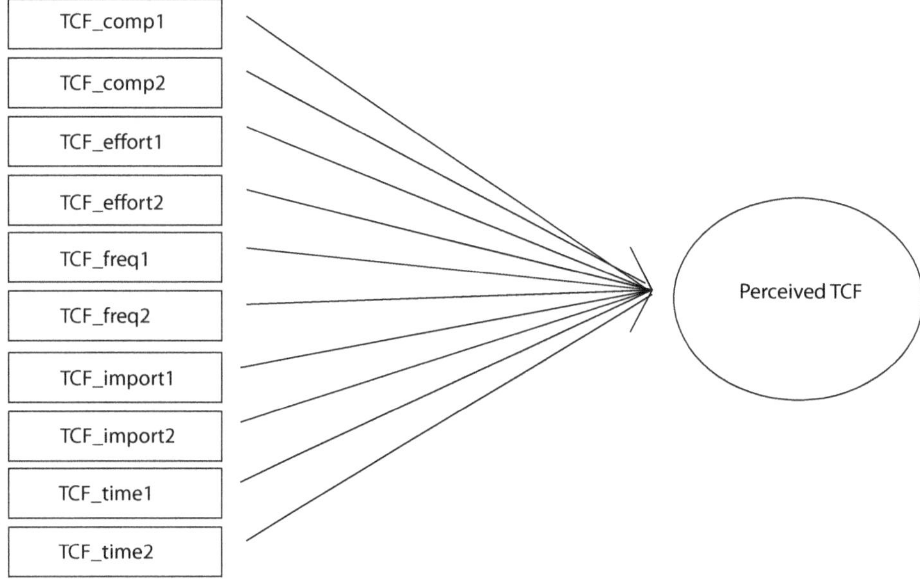

Figure 6.6 Formative Items Measuring the Perceived TCF Construct

Table 6.5 VIF Statistics for Formative Measures—Deviation Scores

Formative item	VIF_loans	VIF_account inquiries
TCF_comp1	1.51	1.40
TCF_comp2	1.67	1.31
TCF_effort1	1.42	1.43
TCF_effort2	1.33	1.23
TCF_import1	1.64	1.39
TCF_import2	1.56	1.25
TCF_time1	1.10	1.20
TCF_time2	1.63	1.24
TCF_freq1	2.83	1.72
TCF_freq2	2.51	1.86

All VIF values (for the financial loan and account inquiry dataset) ranged between 1.1 and 2.8, indicating that multi-collinearity is not present among the formative measures.

Next, the t-values and weights for the formative measures were produced using SmartPLS. Tests of significance were performed using a bootstrap re-sampling procedure.

Table 6.6 Formative Measures, t-values and Item Weights—Deviation Scores

	Financial loans		Account inquiries	
Formative measure	**T-statistic**	**Weight**	**T-statistic**	**Weight**
TCF_comp1 -> TCF	2.69**	0.56	1.73	0.31
TCF_comp2 -> TCF	1.29	0.29	0.28	0.05
TCF_effort1 -> TCF	0.13	0.03	1.46	0.26
TCF_effort2 -> TCF	0.15	0.03	1.99*	0.33
TCF_freq1 -> TCF	0.73	-0.21	2.15*	0.40
TCF_freq2 -> TCF	1.97*	0.47	0.94	-0.21
TCF_import1 -> TCF	0.83	-0.18	0.16	0.04
TCF_import2 -> TCF	0.82	0.16	1.23	0.24
TCF_time1 -> TCF	3.68***	0.61	3.42***	0.49
TCF_time2 -> TCF	1.02	-0.22	0.47	0.09

* significant at $p < 0.05$

** significant at $p < 0.01$

*** significant at $p < 0.001$

Table 6.6 illustrates that only three items showed significant t-statistics (TCF_comp1, TCF_freq2 and TCF_time1) for the financial loan dataset, and three for the account inquiry dataset (TCF_effort2, TCF_freq1 and TCF_time1). In addition, several item weights were relatively low and four item weights were negative.

Cenfetelli and Bassellier (2009, p. 697) recommend that 'if the negatively weighted items are (a) not suppressors or (b) not collinear, they should be included in the remaining analysis and potentially culled over time if they repeatedly behave differently than other indicators'. As shown in Table 6.6, none of the weights was negative in *both* datasets (financial loans and account inquiries). Thus, it seemed reasonable to conclude that no suppressor effects were present. In addition, since multi-collinearity problems were not evident, none of the negative weights was excluded from further analysis.

Finally, the TCF scales were inspected for their portability, or generalisability (Cenfetelli and Bassellier, 2009). Mathieson et al. (2001) suggested linking the formatively assessed construct to a reflectively measured variable *measuring a conceptually equivalent concept.*

Therefore, the inter-construct correlation between the formatively measured TCF construct and its reflectively assessed counterpart was assessed. SmartPLS was used for this purpose. Table 6.7 displays the results.

Table 6.7 Redundancy Analysis—Deviation-Score Analysis

	Financial loans		Account inquiries	
Construct relations	**T-statistic**	**Correlation coefficient**	**T-statistic**	**Correlation coefficient**
TCF (all items) -> TCF reflective	6.89***	0.43	7.68***	0.46
TCF (reduced set of items) -> TCF reflective	6.23***	0.43	6.39***	0.45

* significant at $p < 0.05$

** significant at $p < 0.01$

*** significant at $p < 0.001$

Table 6.7 shows that the formatively measured TCF construct was significantly correlated with the reflective directly measured TCF construct (for both datasets). Cenfetelli and Bassellier (2009) suggested that inter-construct correlation coefficients should exceed a 0.80 threshold. This guideline was provided in reference to the ServQual instrument. This instrument is one of the most validated instruments in the marketing/IS literature. Since the TCF survey questionnaire instrument is newly developed, however, correlation coefficients greater than 0.4 were accepted as adequate.

It is important to note that the inter-construct correlations do not differ significantly when using the reduced set of formative measures (excluding TCF_comp2, TCF_effort1, TCF_import1 and TCF_import2 and TCF_time2). This suggests that the exclusion of these items in future studies when conceptualising the TCF construct with deviation scores would be acceptable.

Discussion

This study employed a parallel-instrument approach to formatively measure the perceived TCF of electronic banking channels. This approach addresses shortcomings of several fit measurement approaches outlined by Venkatraman (1989) including direct fit measurement and fit as co-variation (as explained in the 'Literature Review' section above). Both approaches have been commonly used by IS researchers to assess TTF despite well-known conceptual and analytical issues with them.

The approach illustrated here has several advantages over the fit measurements traditionally employed by TTF researchers. For instance, the parallel-instrument approach advances direct-fit measurement since respondents are not asked to mentally 'calculate' a fit between a given technology (for example, Internet banking) and tasks (for example, account inquiries and mortgages) they perform with it. Instead, individuals are required to answer questions regarding a given variable, A (task-characteristics), and a given variable, B (channel suitability). While the focus of this chapter is on the TCF of electronic banking channels, future research could investigate further tasks and technology combinations (for example, work-related tasks and enterprise resource planning systems) using parallel instruments.

What is more, deviation-score analysis allows researchers to determine a numerically calculated fit between two variables. This advances the fit as co-variation measurement since researchers can numerically quantify fit when investigating statistical correlations with other variables specified in TTF-related research models.

To assess the transportability of the formatively measured TCF construct, a set of reflective TCF measures was created for this study in order to assess consumers' overall perceptions of the task-channel fit. The redundancy analysis showed a strong correlation (approximately 0.45) between both constructs. This suggests that the formatively measured TCF variable captured the most essential aspects of consumers' overall perceptions of the TCF.

Future studies could extend the scope of this research by investigating possible causal relationships between the formatively measured TCF construct and other variables influencing consumers to use electronic banking channels. Following Goodhue and Thompson's (1995) TPC framework, such variables could include expected consequences, affect, social norms, habit and facilitating conditions related to IS (or electronic banking channel) usage.

Likewise, the formatively measured TCF construct could be linked to well-established IS acceptance theories such as TAM (Davis, 1989) and UTAUT (Venkatesh et al., 2003). Dishaw and Strong (1999) have verified a theoretical linkage between TTF and TAM theory. Hence, causal relationships between the formatively measured TCF construct and constructs such as perceived usefulness and ease of use should be investigated in future research.

Another avenue for future research is to apply the TCF measurement approach to different industries deploying similar self-service technologies as banks (for example, the airline industry, supermarkets, and so on). The conceptualised TCF dimensions seem to apply to the activities individuals perform in these use settings as well.

For instance, ordering an airline ticket is often seen as more complicated than checking flight arrival times online (D'Ambra and Wilson, 2004a). Similarly, checking arrival times for aircraft appears to be a time-critical activity whereas consumers usually plan holidays ahead. Likewise, ordering flight tickets includes more effort than checking arrival times since individuals are required to fill out more forms when performing these activities online. These examples outline how the conceptualised TCF dimensions could be reapplied to different e-commerce contexts.

Concluding Comments

The focus of our chapter has been the development of an approach to measuring task-channel fit for electronic banking channels. The TCF was initially conceptualised following extant literature and subsequently reassessed via five exploratory focus group discussions. Next, a parallel instrument was developed using two judgment rounds and two pre-test evaluations. Following this, the instrument was used in a survey; 140 responses for each banking task (account inquiries and financial loans) were gathered. The respondents were New Zealand consumers using Internet banking channels.

An important theoretical contribution of this research is a quantitative assessment of the TCF concept first suggested by Hoehle and Huff (2009). Task-channel fit adapts the task-technology fit theory (Goodhue and Thompson, 1995) to study an online delivery channel rather than a specific technology. To date, no previous study has used TTF theory for this purpose.

Our study also contributes to construct specification and measurement. The TTF theory was originally developed within an organisational context characterised by involuntary use. So far, very little is known about how this concept can be applied at the individual level (Staples and Seddon, 2004). We have addressed this issue by developing and validating a survey questionnaire instrument to measure the TCF of electronic banking channels.

Our third contribution is to banks and financial institutions distributing their products and services through electronic banking channels. Prior to the research reported in this chapter, we conducted relevance checks with several senior managers working for three German banks regarding their perceptions of the TCF concept (Hoehle and Huff, 2009). Those interviews indicated that a measure of task-channel fit would be highly valuable for banking practitioners since it would enable them to better judge which banking products to offer on each of the channels their bank supports (Hoehle and Huff, 2009).

Limitations and Suggestions for Future Research

The data used in the analysis were collected in a university environment. University staff and students are usually technology-savvy and have easy access to computers and the Internet. This might bias the results. A replication of the study drawing from the general population of users of electronic banking channels is essential before the findings can be generalised to a broader audience.

Second, only two tasks (account inquiries and financial loans) were used in combination with Internet banking to test the TCF model. Future research should extend these results by examining TCF in the context of other electronic banking channels (ATMs, phone banking and mobile banking) and other banking tasks.

Third, not all of the formative measure item weights were significant. Further improvement in the TCF measurement items is required to improve the psychometric properties of the TCF construct measure.

References

Bagozzi, R. P. (1982). A field investigation of causal relations among cognitions, affect, intentions and behavior. *Journal of Marketing Research, 19*(4), 562–83.

Baroudi, J. J., Olson, M. H., & Ives, B. (1986). An empirical study of the impact of user involvement on system usage and information satisfaction. *Communications of the ACM, 29*(3) 232–8.

Benbasat, I., Dexter, A. S., & Todd, P. (1986). An experimental program investigating color-enhanced and graphical information presentation: an integration of the findings. *Communications of the ACM, 29*(11), 1094–105.

Bhattacherjee, A. (2001). Understanding information system continuance: an expectation-confirmation model. *MIS Quarterly, 25*(3), 351–70.

Buttermann, G., Germain, R., & Iyer, K. N. S. (2008). Contingency theory 'fit' as gestalt: an application to supply chain management. *Transportation Research Part E: Logistics and Transportation Review, 44*(6), 955–69.

Calder, B. J. (1977). Focus groups and the nature of qualitative marketing research. *Journal of Marketing Research, 14*(3), 353–65.

Campbell, D. J. (1988). Task complexity: a review and analysis. *Academy of Management Review, 13*(1), 40–52.

Cenfetelli, R. T., & Bassellier, G. (2009). Interpretation of formative measurement in information systems research. *MIS Quarterly, 33*(4), 689–707.

Chan, Y. E., Huff, S., Barclay, D. W., & Copeland, D. G. (1997). Business strategic orientation, information systems strategic orientation, and strategic alignment. *Information Systems Research, 8*(2), 125–50.

Conrad, C. A., Brown, G., & Harmon, H. A. (1997). Customer satisfaction and corporate culture: a profile deviation analysis of a relationship marketing outcome. *Psychology and Marketing, 14*(7), 663–74.

Cooper, R. B., & Zmud, R. (1990). Information technology implementation research: a technological diffusion approach. *Management Science, 36*(2), 123–39.

D'Ambra, J., & Wilson, C. S. (2004a). Explaining perceived performance of the World Wide Web: uncertainty and the task-technology fit model. *Internet Research, 1*(4), 294–310.

D'Ambra, J., & Wilson, C. S. (2004b). Use of the World Wide Web for international travel: integrating the construct of uncertainty in information seeking and the task-technology fit (TTF) model. *Journal of the American Society for Information Science and Technology, 55*(8), 731–42.

Davis, F. D. (1989). Perceived usefulness, perceived ease of use, and user acceptance of information technology. *MIS Quarterly, 13*(3), 319–39.

Davis, F. D., Bagozzi, R. P., & Warshaw, P. R. (1989). User acceptance of computer technology: a comparison of two theoretical models. *Management Science, 35*(8), 982–1002.

DeLone, W. H., & McLean, E. R. (1992). Information systems success: the quest for the dependent variable. *Information Systems Research, 3*(1), 60–95.

Deutsche Bank Research. (2006). *DB research: banking online boosts and curbs customer loyalty*. Research Report, Deutsche Bank Research, Frankfurt am Main, Germany. Retrieved from <http://www.dbresearch.de/>

Diamantopoulos, A., & Winklhofer, H. M. (2001). Index construction with formative indicators: an alternative to scale development. *Journal of Marketing Research, 38*(2), 269–77.

Diamantopoulos, A., Riefler, P., & Roth, K. P. (2008). Advancing formative measurement models. *Journal of Business Research, 61*(12), 1203–18.

Dickson, G. W., DeSanctis, G., & McBride, D. J. (1986). Understanding the effectiveness of computer graphics for decision support: a cumulative experimental approach. *Communications of the ACM, 29*(1), 40–7.

Dishaw, M. T., & Strong, D. M. (1999). Extending the technology acceptance model with task-technology fit constructs. *Information & Management, 36*(1), 9–21.

Edwards, J. E. (2001). Ten difference score myths. *Organizational Research Methods, 4*(3), 265–87.

Evermann, J., & Tate, M. (2009). Constructs in the mist: the lost world of the IT artifact. *Proceedings of the International Conference on Information Systems (ICIS)*, Phoenix, Ariz.

Ferratt, T. W., & Vlahos, G. E. (1998). An investigation of task-technology fit for managers in Greece and the US. *European Journal of Information Systems, 7*(2), 123–36.

Fishbein, M., & Ajzen, I. (1975). *Belief, Attitude, Intention and Behavior: An introduction to theory and research.* Reading, Mass.: Addison-Wesley.

Forrester Research. (2007). *European Mobile Banking: An inconvenient truth*, 25 October, Forrester Research, Cambridge, Mass.

Gattiker, T. F., & Goodhue, D. L. (2005). What happens after ERP implementation: understanding the impact of inter-dependence and differentiation on plant-level outcomes. *MIS Quarterly, 29*(3), 559–85.

Goetz, O., Liehr-Gobbers, K., & Krafft, M. (2010). Evaluation of structural equation models using the partial least squares (PLS) approach. In V. E. Vinnzi, W. W. Chin, J. Henseler & H. Wang (eds), *Handbook of Partial Least Squares* (pp. 691–712). Heidelberg, Germany: Springer.

Goodhue, D. L. (1995). Understanding user evaluations of information systems. *Management Science, 41*(12), 1827–44.

Goodhue, D. L. (1998). Development and measurement validity of a task-technology fit instrument for user evaluations of information systems. *Decision Sciences, 29*(1), 105–38.

Goodhue, D. L., & Thompson, R. L. (1995). Task technology fit and individual performance. *MIS Quarterly, 19*(2), 213–36.

Hoehle, H., & Huff, S. (2009). Electronic banking channels and task-channel fit. *Proceedings of the International Conference on Information Systems (ICIS 2009)*, Phoenix, Ariz.

Jarvenpaa, S. L. (1989). The effect of task demands and graphical format on information processing strategies. *Management Science, 35*(3), 285–303.

Jenkins, C. D., Zyanski, S. J., & Roseman, R. H. (1971). Progress toward validation of a computer-scored test for the Type A coronary-prone behavior pattern. *Psychometric Medicine, 33*(3), 193–202.

Jiang, J. J., Klein, G., & Carr, C. L. (2002). Measuring information system service quality: ServQual from the other side. *MIS Quarterly, 26*(2), 145–66.

Junglas, I., Abraham, C., & Ives, B. (2009). Mobile technology at the frontlines of patient care: understanding fit and human drives in utilization decisions and performance. *Decision Support Systems, 46*(3), 634–47.

Karimi, J., Somers, T. M., & Gupta, Y. P. (2004). Impact of environmental uncertainty and task characteristics on user satisfaction with data. *Information Systems Research, 15*(2), 175–93.

Keen, P. (1980). MIS research: reference disciplines and a cumulative tradition. *Proceedings of the First International Conference on Information Systems,* Philadelphia, Pa, 9–18.

Kleijnen, M., Wetzels, M., & de Ruyter, K. (2004). Consumer acceptance of wireless finance. *Journal of Financial Services Marketing, 8*(3), 206–17.

Klein, G., Jiang, J. J., & Cheney, P. (2009). Resolving difference score issues in information systems research. *MIS Quarterly, 3*(4), 811–26.

Klopping, I. M., & McKinney, E. (2004). Extending the technology acceptance model and the task-technology fit model to consumer e-commerce. *Information Technology, Learning, and Performance Journal, 22*(1), 35–48.

Kristof, A. L. (1996). Person-organization fit: an integrative review of its conceptualization, measurement, and implications. *Personnel Psychology, 49*(1), 1–49.

Landry, F. J., Rastegary, H., Thayer, J., & Colvin, C. (1991). Time urgency: the construct and its measurement. *Journal of Applied Psychology, 76*(5), 644–57.

Lefebvre, E., Lefebvre, L. A., & Prefonta, L. (1997). Technological learning and organizational context: fit and performance in SMES. *Proceedings of the 30th Hawaii International Conference on System Sciences (HICSS), Wailea, HI, 3,* 464–74.

Liao, Z., & Cheung, M. T. (2002). Internet-based e-banking and consumer attitudes: an empirical study. *Information & Management, 39*(4), 283–95.

Limayem, M., Hirt, S. G., & Cheung, C. M. K. (2007). How habit limits the predictive power of intention: the case of information systems continuance. *MIS Quarterly, 31*(4), 705–37.

Lin, C. S., Wu, S., & Tsai, R. J. (2005). Integrating perceived playfulness into expectation-confirmation model for web portal context. *Information & Management, 42*(5), 683–93.

McKinney, V., Yoon, K., & Zahedi, F. (2002). The measurement of web-customer satisfaction: an expectation and disconfirmation approach. *Information Systems Research, 13*(3), 296–315.

Mathieson, K., Peacock, E. & Chin, W. W. (2001). Extending the technology acceptance model: the influence of perceived user resources. *Database for Advances in Information Systems, 32*(3), 86–112.

Miles, M. B., & Huberman, A. M. (1994). *An Expanded Sourcebook-Qualitative Data Analysis*. Thousand Oaks, Calif.: Sage Publications.

Mitchell, V., Mohammad, H., Stock, T., & Wei, X. (2007). Fit perspectives and theory building in information systems. *Proceedings of the 40th Annual Hawaii International Conference on System Sciences (HICSS), Waikoloa, HI,* 1530–605.

Nadkarni, S., & Gupta, R. (2007). A task-based model of perceived website complexity. *MIS Quarterly, 31*(3), 501–24.

Ortiz de Guinea, A., & Markus, M. L. (2009). Why break the habit of a lifetime? Rethinking the roles of intention, habit, and emotion in continuing information technology use. *MIS Quarterly, 33*(3), 433–44.

Park, C. W., Im, G., & Keil, M. (2008). Overcoming the mum effect in IT project reporting: impacts of fault responsibility and time urgency. *Journal of the Association for Information Systems, 9*(7), 409–31.

Parker, S. C., & van Witteloostuijn, A. (2010). A general framework for estimating multidimensional contingency fit. *Organization Science, 21*(2), 540–53.

Petter, S., Straub, D. W., & Rai, A. (2007). Specifying formative constructs in information systems research. *MIS Quarterly, 31*(4), 623–56.

Prescott, J. E. (1986). Environments as moderators of the relationship between strategy and performance. *Academy of Management Journal, 29*(2), 329–46.

Rangan, V. K., Menezes, M. A. J., & Maier, E. P. (1992). Channel selection for new industrial products: a framework, method, and application. *Journal of Marketing, 56*(3), 69–82.

Reinsch, N. L., & Beswick, R. W. (1990). Voice mail versus conventional channels: a cost minimization analysis of individuals' preferences. *Academy of Management Journal, 33*(4), 801–16.

Robey, D. (1979). User attitudes and management information system use. *Academy of Management Journal, 22*(3), 527–38.

Sabberwal, R., & Chan, Y. E. (2001). Alignment between business and IS strategies: a study of prospectors, analyzers, and defenders. *Information Systems Research, 12*(1), 11–34.

Sayar, C., & Wolfe, S. (2007). Internet banking market performance: Turkey versus the UK. *The International Journal of Bank Marketing, 25*(3), 122–41.

Shirani, A. I., Tafti, M. H. A., & Affisco, J. F. (1999). Task and technology fit: a comparison of two technologies for synchronous and asynchronous group communication. *Information and Management, 36*(3), 139–50.

Staples, D. S., & Seddon, P. (2004). Testing the technology-to-performance chain model. *Journal of Organizational and End User Computing, 16*(4), 17–36.

Strauss, A. L., & Corbin, J. (1998). *Basics of Qualitative Research: Techniques and procedures for developing grounded theory*. [Second edn]. Thousand Oaks, Calif.: Sage Publications.

Swanson, E. B. (1987). Information channel disposition and use. *Decision Sciences, 18*(1), 131–45.

Tan, M., & Thompson, S. H. (2000). Factors influencing the adoption of internet banking. *Journal of the Association of Information Systems, 1*(1), 1–42.

Tesch, D., Jiang, J. J., & Klein, G. (2003). The impact of information system personnel skill discrepancies on stakeholder satisfaction. *Decision Sciences, 34*(1), 107–29.

Thatcher, S. (2001). The mediating role of identity fit: understanding the relationship between communication media, demographic differences, and creativity. *Proceedings of the 34th Annual Hawaii International Conference on System Sciences (HICSS), Waikoloa, HI.*

Tornatzky, L. G., & Klein, K. J. (1982). Innovation characteristics and innovation adoption-implementation: a meta-analysis of findings. *IEEE Transactions on Engineering Management, 29*(1), 28–45.

Venkatesh, V., & Goyal, S. (2010). Expectation disconfirmation and technology adoption: polynomial modeling and response surface analysis. *MIS Quarterly, 34*(2), 281–303.

Venkatesh, V., Morris, M. G., Davis, G. B., & Davis, F. D. (2003). User acceptance of information technology: toward a unified view. *MIS Quarterly, 27*(3), 425–78.

Venkatraman, N. (1989). The concept of fit in strategy research: toward verbal and statistical correspondence. *The Academy of Management Review, 14*(3), 423–44.

Vessey, I. (1991). Cognitive fit: a theory-based analysis of the graphs versus tables literature. *Decision Sciences, 22*(2), 219–41.

Wood, R. E. (1986). Task complexity: definition of a construct. *Organizational Behavior and Human Decision Processes, 31*(1), 60–82.

Yuan, Y., Archer, N., Conelly, C. E., & Zheng, W. (2010). Identifying the ideal fit between mobile work and mobile work support. *Information and Management, 47*(3), 125–37.

Zeithaml, V. A., Parasuraman, A., & Berry, L. L. (1990). *Delivering Quality Service: Balancing customer perceptions and expectations.* New York, NY: The Free Press.

Zigurs, I., & Buckland, B. K. (1998). A theory of task/technology fit and group support systems effectiveness. *MIS Quarterly, 22*(3), 313–34.

Zigurs, I., Buckland, B. K., Connolly, J., & Wilson, E. V. (1999). A test of task-technology fit theory for group support systems. *Data Base for Advances in Information Systems, 30*(3–4), 34–50.Table 6.3 Items Used in Constructing the Construct Measures

7. Theory Building in Action: Structured-case with action interventions

Gerlinde Koeglreiter
University of Melbourne

Ross Smith
RMIT University

Luba Torlina
Deakin University

Abstract

In this chapter an enhancement of the structured-case approach, including action research-style interventions within structured-case cycles, is discussed. An exemplar of this approach is provided through a knowledge management study. The major benefits of this approach include the flexibility of the resulting research process to include cycles both with and without action interventions, its capacity to support a study involving many research cycles conducted over an extended time scale (amounting, in the study described, to several years) and, most importantly, the capacity of the approach to make visible the theory building that takes place.

Introduction

This chapter discusses recent developments in the application of the structured-case (SC) approach of Carroll and Swatman (Carroll, 2000; Carroll and Swatman, 2000). Specifically, a theory-building approach using action research (AR) interventions as part of a larger program of research undertaken in an SC mode is presented.

Structured-case and AR share characteristics that facilitate their combined use, including the iterative conduct of research cycles that are structured into phases of planning, data collection, analysis and reflection. Further, both methodologies support research conducted within the interpretive research paradigm.

For the study described in this chapter, a methodology was required that would support a complex research agenda, including investigation of phenomena drawing on rich field data and extensive literature from distinct disciplines, sensitive stakeholder interactions and a study of dynamics of these interactions. These factors might be relevant to many projects in the multifaceted information systems (IS) discipline, thus making the proposed combination of research methods beneficial to many research projects facing similar research settings. Further, the proposed combined methodology has been found effective for a project that seeks to address both theoretical requirements and practical goals, including responsiveness to research participants' needs, action interventions and theory building over an extended period (amounting, in the present study, to several years).

In this particular context, pure AR might pose challenges to theory building. Action research requires an intervention to serve the dual imperative of inducing change to improve an organisational problem situation and building theory (McKay and Marshall, 2001; McKay and Marshall, 2007). In practice, however, to accommodate organisational needs, multiple intervention cycles over a short period might require intense periods of research (Schultze, 2000), often leaving the theory-building part behind to occur at a later stage.

The research settings discussed above create the need for a research methodology that combines two forms of research activity in an iterative process. This chapter argues that a structured approach, as facilitated by SC, with reduced pressure on inducing change as mandated by AR, but with the advantages of the rich information obtained in participant observation and evaluation of organisational impact, allows the researcher to spend more time on reflection, increasing the depth of theory being built. The focus on the theoretical foundation facilitates careful consideration of subsequent interventions, so aligning the practical response with a coherent theoretical framework. This approach might therefore find wider applicability to IS research projects that pursue systems-based problem solving both for theory building and for testing.

Structured-case has been promoted as a means of supporting theory building as part of case study research (Carroll, 2000; Carroll and Swatman, 2000). While SC follows a cyclical approach, as does AR, it focuses on theory building based on analysis of the extant literature and reflection on the present and previous cycles.

Structured-case might be extended to include interventions involving participant/participatory observation and evaluation of organisational impact to obtain rich data and develop, deepen and test theory in a cyclical approach. In the present chapter, a study is outlined that applies such an approach. This approach was found particularly suitable for a project with a complex agenda, conducted over a long period and seeking to achieve both change management and rich theory building.

Specifically, this chapter reports methodological findings from a PhD project that combined SC and AR interventions to develop a multifaceted framework of organisational knowledge management (KM). The chapter is structured as follows: the background section sets the research context of the study and describes the problem situation in terms of the requirements that were placed upon the selection of an appropriate methodology. Subsequent sections provide an introduction to AR and structured-case. Then, the findings are reported, bringing together the elements of these two methodologies in a coherent research program and describing the process of analysis and theory building that was conducted over four cycles of research. The concluding section provides a brief discussion of the perceived advantages of the process employed and summarises the chapter's contribution to research practice.

Background

The research project described in this chapter sought to build a multifaceted framework for organisational KM involving a community of practice (CoP), as the core group to be studied, and other entities of the wider organisation in which the CoP existed, as well as interdependencies between the parties.

The program of research involved investigation of the role of a CoP in organisational KM and was conceived as a 'bottom-up' study. The CoP was a group of academics teaching systems implementation subjects in the School of Information Systems at an Australian university. More specifically, the research sought to understand the role a CoP might play in collaborative KM and knowledge strategy development.

Extant research has acknowledged that CoPs might be viewed as a bottom-up approach to KM, and that such approaches need to be coordinated with top-down KM efforts (Wenger et al., 2002). The research aimed to provide insights into how these two approaches might be implemented as part of a coherent, comprehensive KM strategy. A comprehensive KM strategy would be one that integrates top-down KM (strategic and mandated by management) and bottom-up KM (work practice-oriented and owned by a CoP at lower hierarchical levels in the organisation). This approach considers a range of factors related to knowledge work, performed both by CoPs and by management, and entails a strong social aspect, involving knowledge flows between a CoP and the wider organisation.

Consistent with the aim and nature of the project, the research agenda included exploration of complex and interlacing considerations of the social aspects that facilitate a CoP, the nature of knowledge work and knowledge work support, both technical and non-technical. The research also included the organisational elements of management and leadership style, strategic alignment and disciplinary and organisational boundary conflicts. Importantly, a bottom-up approach was taken to investigate KM, focusing primarily on inputs from staff at lower hierarchical levels of the organisation.

Initially, the aim was predominantly practice oriented since the researcher saw the need, and opportunities, to improve the situation of the community of practice. This kind of situation can be considered as ideal for an AR project aiming at significant change. In the early stages of the project, however, the need for adjustments to the research process was recognised. Factors contributing to the changes in the research project involved organisational changes impacting on the priorities of the group under study, and an opportunity for conducting observations over a longer period to gain a deeper understanding of the CoP knowledge work. As a result, the opportunity was taken to develop the project in the direction of a more complex research agenda, shifting the primary goal of introducing organisational change, as practiced in AR, to generating knowledge (Gaventa and Cornwall, 2001). The goal had thus shifted to theory building as the primary purpose of the study, introducing only limited organisational change.

Action research interventions focus primarily on problem solving, with a secondary focus on the generation of knowledge (Gaventa and Cornwall, 2001). Structured-case, as the overarching methodological framework, involves cycles of interpretive case study, with a focus on theory development drawing upon field data. Structured-case does not intend directly to induce change and address the organisational issues faced by the CoP, but rather seeks to develop deeper understanding of the systemic aspects of knowledge work conducted by the group. Both methods were supposed to inform each other within a single research design, enriching a series of conceptual frameworks. A further requirement of the adopted methodology was that it should facilitate the extraction of theory from masses of field data and broad research themes (Carroll and Swatman, 2000).

The requirements described above can be accommodated within the SC approach, which originally drew upon the AR process, and which allows selection of any research processes, tools and techniques at each cycle of activity for theory building (Carroll and Swatman, 2000).

This chapter acknowledges similarities to the action case method (Vidgen and Braa, 1997) that combines soft case study and AR as a means of addressing the need to balance intervention and interpretation. Vidgen and Braa (1997) recommend action case for projects of short to medium duration, small-scale interventions and a narrow research focus. Action case was evaluated as an alternative method in our situation, but was found unsuited to the planned long-term research project, with its flexible and hence complex agenda and its strong focus on a transparent presentation of theory building throughout the project.

In response to these imperatives, an approach was adopted that included AR interventions within a broader program of structured-case. In the following sections the elements of SC and AR are discussed in terms of their contribution to the overarching research program.

Structured-Case

Structured-case (SC) is 'a methodological framework that assists IS researchers to undertake and assess theory building research within the interpretive paradigm, and explains its value in achieving convincing explanations that are strongly linked to both the research themes and data collected in the field' (Carroll and Swatman, 2000, p. 235). Structured-case supports the use of different tools and processes to build and validate theory.

The SC methodology builds on the case study methodology and employs structural elements of AR (Carroll, 2000). Its basis is the case, in the sense of what is being studied. 'Structured' refers to the formal process model, and consists of three elements. The first element is a conceptual framework representing the researcher's aims. Structured-case suggests that the research subject matter (Carroll, 2000) can be based on either an assumption that concepts will emerge purely from large amounts of data collected, with little predefined structure, as advocated in grounded theory (Glaser, 1992; Strauss and Corbin, 1998), or on preconceived notions and a conceptual structure that can underpin the research, based on available, but possibly scarce, resources (Orlikowski and Baroudi, 1991). Second, a predefined research cycle guides data collection, analysis and interpretation. Finally, literature-based scrutiny compares and contrasts the outcomes of the research process with a broad range of literature to support or challenge the theory built (Carroll and Swatman, 2000).

In SC, a multi-staged process helps the researcher to organise data collection and analysis (Carroll, 2000), providing a predefined structure for the conduct of research and development of theory. The SC research cycle includes stages of planning, data collection, analysis and reflection (see Figure 7.1).

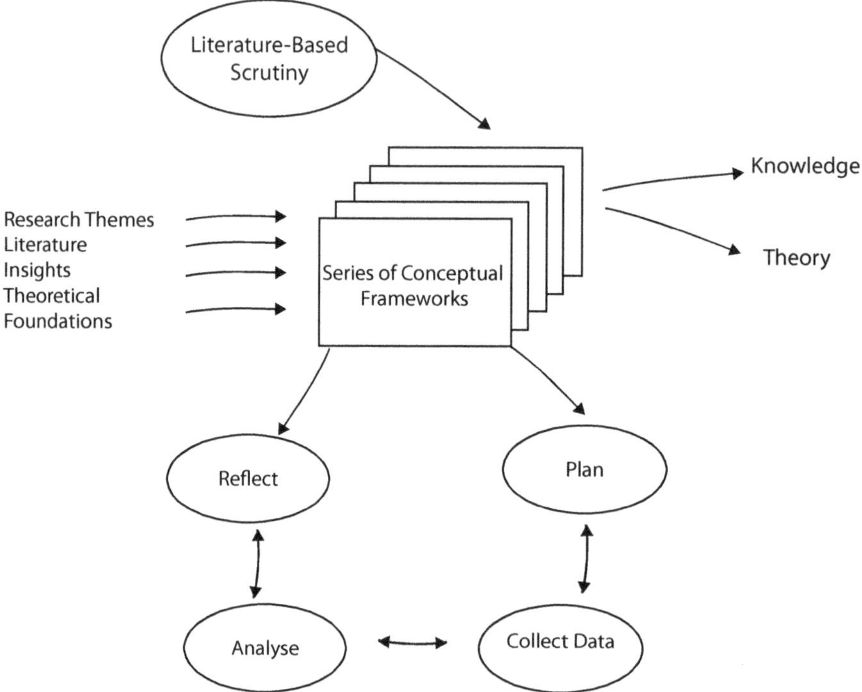

Figure 7.1 The Structured-Case Research Cycle

Source: Carroll and Swatman (2000).

The development of the conceptual framework is conducted through a series of research cycles. Each of the cycles informs and extends the previous cycle, refining the conceptual framework to a point of saturation, which is determined by the researcher. The reflective stages as part of the cyclical approach also support a process of abstraction. Structured-case supports theory building within the interpretative paradigm and assists in generating a significant level of abstraction.

Action Research

Action research (AR) has been defined as 'a cognitive process that depends on social interaction between the observers and those in their surroundings' (Baskerville and Wood-Harper, 1998, p. 92).

What distinguishes AR from other methodologies is that it involves practical problem solving that has theoretical relevance (Mumford, 2001). Action research is conducted in cycles of interventions, where outcomes examined in one cycle are the input to the next cycle, with the overall intent to change the situation for the better. The researcher is heavily involved in the organisational life of the research subjects and might even be one of their colleagues.

Essentially, AR consists of two major stages (see Figure 7.2): a diagnostic stage where a social situation is collaboratively analysed; and a therapeutic stage involving collaborative change experiments where changes are introduced and the effects studied (Blum, 1955). These two stages are implemented in a cyclical process, linking theory and practice (Baskerville and Wood-Harper, 1996).

Baskerville and Wood-Harper (1998) identify four characteristics by which the various AR methods can be classified: process model, structure, typical researcher involvement and primary goals. The AR process model might be reflective (theory-in-use versus espoused theory), linear (single sequence of activities) or iterative (repeated cycles of activity until a satisfactory outcome is achieved). The structure of AR can be rigorous, following predefined rules, or fluid, where activities are defined loosely. The typical researcher involvement can be facilitative (researcher takes an advisory role), expert (the researcher's task is to solve the problem) or collaborative (researcher and participants are equal).

If applied in organisational development, AR aims at improving the human organisation, including the development of the social conditions of the organisation. Action research, as used in systems design, seeks to create or modify organisational systems. If undertaken with the educational goal of creating scientific knowledge, AR attempts to produce a generalisable understanding that practitioners can use in different settings or that other researchers build upon in subsequent studies.

To ensure that both the problem-solving and the research interests are addressed, a parallel dual-cycle process that also addresses the research interest has to be followed (McKay and Marshall, 2001). Eden and Huxham (1995) suggest that this should take place via a comprehensive AR design, which involves a continuous writing process to inform theory exploration and implicit pre-understanding.

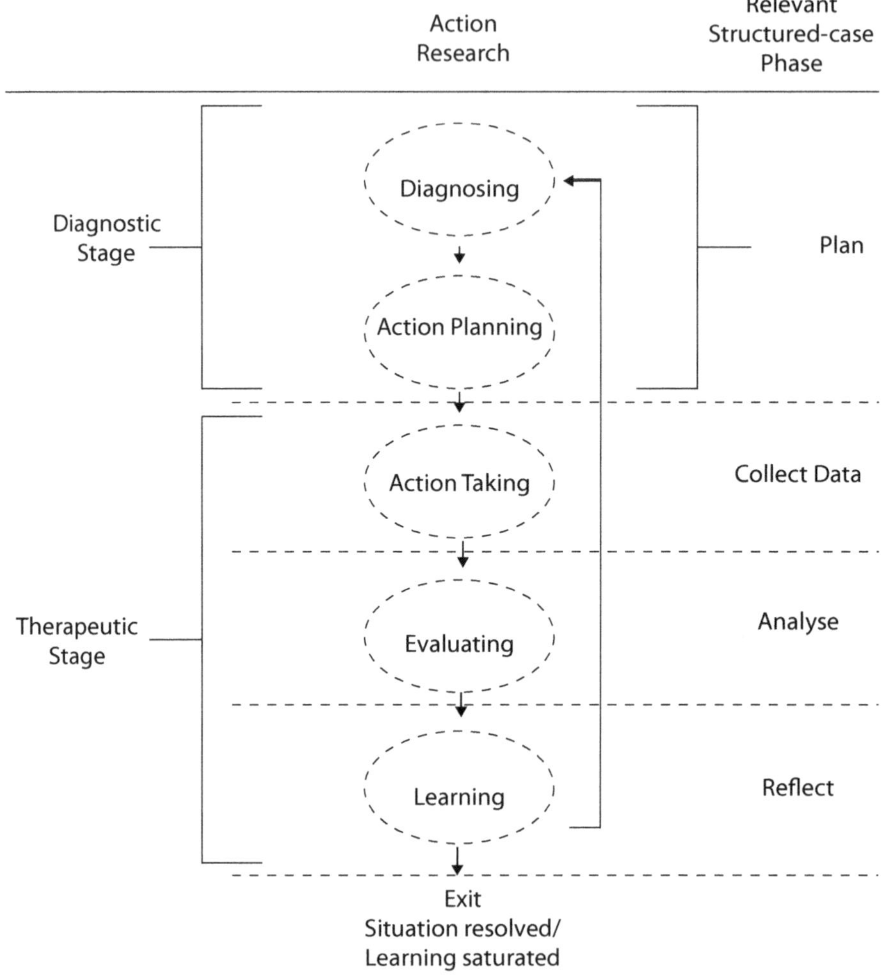

Figure 7.2 The Action Research Cycle

Source: Adapted from Baskerville and Wood-Harper (1996, 1998); and Blum (1955).

Structured-Case with Action Interventions

As already noted, the SC approach was adapted in the present study to include AR-style interventions within three of four research cycles. As such, practice-oriented change took place, but within a SC framework that facilitated the emergence of theoretical insights. One research cycle did not include action in the form of an intervention with the aim to introduce change, but took a more conventional case study approach, to investigate the theoretical linkage of the CoP's practice with the formal organisation. Reporting the research undertaken within a SC framework accommodates this. Figure 7.3 depicts the modified research cycle including action interventions, encompassing a synthesis of the process and essence of SC (see Figure 7.1) and AR (see Figure 7.2).

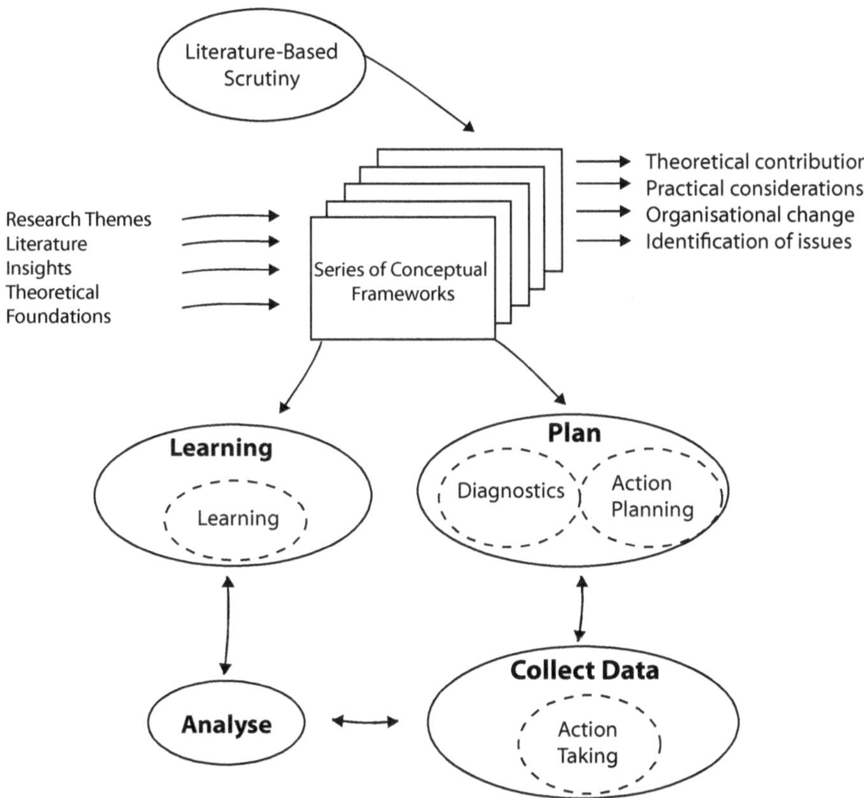

Figure 7.3 The Structured-Case Research Approach Including Elements from AR

Source: Adapted from Baskerville (1999); and Carroll (2000).

The following sections describe each stage in the research approach, depicted in Figure 7.3, in the context of SC, action interventions and their application in the present study.

Plan

Both SC and AR cycles involve a planning phase. In SC the planning stage includes formulation of the research themes extracted through theory examination and data collection planning, including considerations of data collection requirements and data analysis techniques (Yin, 1994), as well as selection of the research site and participants.

Action research starts with a diagnostic phase involving considerations of practical requirements, such as identifying primary problems as the underlying causes of the organisation's desire to change (Baskerville, 1999). In AR the planning phase is primarily concerned with action or intervention planning to relieve or improve primary problems identified in the diagnostic phase. Action is planned through guidance of the theoretical framework indicating the desired future state and changes that would achieve such a state (Baskerville, 1999).

In the present study, the AR diagnostic stage was integrated into the SC planning phase to address the requirements and issues encountered by the research subjects, setting a practical objective for each research cycle. Then, literature considered relevant for the diagnosed problem situation was analysed and used as a basis for developing theory-oriented interview questions. Data collection planning primarily involved the design of interview schedules. For cycles with action interventions, workshop agendas were agreed with participants.

Because not all cycles of research need to include action interventions, the integration of the diagnostic and action planning stages is optional and presents a sub-structure to the overall SC cycle structure. While logically intertwined, the practical and theoretical components of the planning phase are, therefore, clearly separated in the presentation of action and theory input.

Collecting Data

In AR, data collection is undertaken in an 'action taking' phase primarily involving notes from participants or participatory observation (Jorgensen, 1989; Kemmis and McTaggert, 2000). This 'data collection in action' approach has exposed AR to criticism as simply a consultation exercise masquerading as research and hence as lacking rigour (Baskerville and Wood-Harper, 1996).

In SC, data collection is guided by the plan devised in the planning stage. By adhering to the SC cycle structure, the focus on theory is continually revisited throughout the stages of the cycle. Data collection and analysis may be overlapping, as immediate analysis of field notes containing the researcher's interpretations might open up new areas of exploration (Carroll, 2000). The data collection process, therefore, involves adjustments responding to opportunities, unexpected outcomes and emergent themes (Carroll, 2000).

Where AR-style interventions were conducted, the action-taking phase involved implementation of the planned action in the form of workshops. Further, interview schedules were designed to contain questions that reflected participants' impressions of the intervention sessions, relating those with concepts identified in the theory examination stage. While all workshop participants were asked the same questions in the same order, the questions were designed to allow the interviewee to describe specific situations as examples. Adjustments in the data collection phase involved addition and refinement of interview questions between interviews.

The collection of data in two formats (workshops and interviews) enabled a separation of the practical and theoretical focuses. Workshops were conducted in a semi-structured manner with sufficient flexibility to address the diagnosed practical problem. Further, this fluid style of action taking is conducive to generating emergent themes through participants raising issues that might not be directly related to the problem addressed in that cycle, but are nevertheless of interest to the researcher in subsequent research cycles or to enrich the understanding of complex situations when presenting an account. The interview format enforced a stronger theory orientation and provided participants with an opportunity for deep reflection, both on the practical situation and in relation to theoretical concepts that were integrated into the interview questions.

Analysis

While the analysis phase may involve the same techniques in both AR and SC, the strong theoretical focus of SC demands a stronger guidance of analysis by literature.

Analysis involves an iterative process of reading and rereading the vast amounts of raw data typical of qualitative research, developing a deep understanding and relating the data to the conceptual framework (Carroll, 2000). Techniques involve coding related to research themes from the conceptual framework and the identification of new concepts based on themes emerging in the course of analysis.

In the present study, data were analysed in iterations. First came the coding of data based on concepts identified in the literature and the evolving framework. Second was the identification of new concepts through a process of writing up interview summaries and extraction of concepts that the research identified as new. Third, a microanalysis was undertaken to examine the underlying meaning of the text and to extract more concepts. Fourth, concepts were interlinked into higher level categories.

While both SC and AR involve a data analysis or evaluation stage, SC advocates this stage as a central element of the research, acknowledging that analysis is undertaken in a nonlinear fashion and may occur throughout the various stages of one or multiple SC cycles (Carroll 2000). For example, in the present study the researchers returned to transcripts of earlier cycles to reanalyse data in the context of the current cycle, or to identify concepts and issues that spanned cycles, so informing the emerging theory on a higher level. This assisted in the development of a coherent framework that addressed issues from a diverse set of angles.

Reflection

Both SC and AR include a reflection stage in their research cycle structure. In SC, deliberate reflection and critical analysis of any interpretations are formal stages of the research process, derived from AR (Carroll, 2000). In AR, reflection involves an evaluation step to assess practical and theoretical outcomes and to critically consider the influences of the intervention on the outcomes (Baskerville, 1999). Reflection in AR also includes the formulation of learning, where new knowledge gained during the intervention flows into the organisation, or alternatively triggers a new cycle if the outcomes are considered unsuccessful or new issues are identified (Baskerville, 1999).

In integrating action interventions in SC cycles, it is essential that the learning stage from AR is also integrated. In the present study, the learning stage assisted in identifying new issues that were addressed in subsequent research cycles, so creating a 'practical' double loop, in parallel with the theory-focused SC loop, which feeds directly into the diagnostic stage of the planning phase in a new cycle of research.

The reflection phase focuses on theory building based on the understanding of theory as a system of interconnected ideas that condenses and organises knowledge (Neuman, 2006). Theorising involves relating the findings to outcomes of previous research cycles, revisiting literature (Carroll, 2000) or returning to informants to confirm tentative interpretations (Trauth, 1997). The reflection stage entails iteration between data (from current and previous cycles), the tentative findings and the inputs to the conceptual framework and recording of the rationale for changing the conceptual framework (Carroll, 2000). Outcomes of reflection include challenge and support of the conceptual framework, or revision and update, based on the findings of the current research cycle. The result is an extended conceptual framework incorporating new concepts and/or refined existing concepts.

In summary, the enhancement of SC with action interventions strengthens the evolving conceptual framework through the parallel presentation of the theoretical contribution and immediate testing of the framework through practical considerations, organisational change and potential identification of issues.

An Application of Structured-Case with Action Interventions

As discussed above, the research approach taken in the present study sought to link the SC concepts of Carroll and Swatman (2000) (see Figure 7.1) with the AR notion of diagnostic and therapeutic stages (Figure 7.2). Specifically, the research adopted the SC method with action interventions and was conducted in four cycles (see Table 7.1) to investigate the CoP as the 'case' and its interaction with the wider organisation in a KM context. Each SC cycle typically consists of stages of planning, data collection, analysis and reflection. Some of these phases were broken down further to address elements of AR, including diagnostics in the planning phase, action taking in the data collection phase and learning as part of the reflection phase. As required for SC, the findings were captured in an evolving conceptual framework (CF1–CF4), visually representing the theory being built.

Action interventions were undertaken as a series of group workshops (three workshops, two hours each), involving a CoP of five lecturers in IS and selected members of the wider organisation. These interventions were followed up in reflective interviews (semi-structured and lasting one hour). The series of interventions evolved, following the needs of the group as extracted in the analysis phase at the end of each research cycle. Theory building followed the set theme of bottom-up KM influences observed in the interaction of the CoP with the wider organisation.

The combination of SC with action interventions provided an opportunity to facilitate change and to undertake the research and theory building. This mixed method removed some of the rigidity associated with single methodologies and hence allowed for flexibility. For example, in the present study one cycle did not involve any action intervention (CF2), and a follow-up investigation of change was undertaken in only one of the four research cycles (CF3). As such, the four cycles would not have formally met the requirements of an AR project.

As an example, to assist readers in understanding Table 7.1, a brief description of research cycle one is provided in what follows. Research cycle one involved an action intervention that aimed at declaring a group as a community of practice. The *planning phase* started with a *diagnosis* that the group required a focused environment to exchange information on their work-related projects and identify common interests, and that individuals understand the concept of a CoP and identify themselves as CoP members. To underpin this practical goal, relevant *theory* on CoP characteristics was examined. Finally, in *action planning*, a workshop was planned to address the practical goals and a follow-up interview schedule was designed to bring together reflections on the workshop in conjunction with the theory examined. The *data collection* phase involved *action* in the form of a workshop, where CoP members presented their work and engaged in conversations on each other's work. Data collection *methods* included workshop observations that were logged by the researcher and reflective interviews with individuals. Following transcription of the interviews, data were *analysed* based on the concepts identified in the theory embedded in the interview schedule as well as issues and themes emerging from the group discussion and individual reflection. In the *reflection* phase the initial diagnosis was revisited and it was *concluded* that the goal had been achieved.

As required in SC, components that represent and describe theoretical and emerging *concepts* were captured in the evolving conceptual framework (CF1), including identity of the CoP, perceptions of organisational management, the relationship between the CoP and management, and knowledge work. *Learning* from reflection on the practical outcomes of the cycle was identified in that the CoP maintains complex and in part problematic relationships with entities of the wider organisation. This was addressed and investigated further in a subsequent research cycle (cycle three).

Conclusion

The integration of SC and AR as applied in the reported study brought with it a number of benefits that might be useful for other IS research projects.

First, the SC/AR integrated approach serves the purpose of developing and testing a conceptual framework in an iterative process. The modified methodology suits the research agenda of theory building by looking at the organisational situation and the research participants through different lenses. From a practical perspective, action interventions aid improvement of the organisational situation.

Second, with action interventions being optional in the proposed approach, a research cycle that focuses on purely theoretical aspects can be included. This can be seen as different to a mixed-method approach since the structure of the presentation of the research process and the development of the conceptual framework remain consistent with the presentation of the other research cycles.

Third, the primary focus on supporting theory building reduces the expectation of achieving substantial organisational change inherent in the AR approach. This might be considered useful by those employing action interventions for the first time or for researchers who lack the organisational power or stakeholder support that is generally required to induce and evaluate significant change.

Table 7.1 Overview of Cycles Based on Structured-Case Phases

Phase/Cycle		Cycle one (CF1)	Cycle two (CF2)	Cycle three (CF3)	Cycle four (CF4)
Plan	Diagnose	Identify CoP activities and declare membership.	n.a.	Boundary conflict. Mediate between CoP and wider organisation.	Stagnation of CoP as identified in CF1. Identify new direction of CoP.
	Theory	Examine theory on CoP characteristics to be able to declare identified group as CoP.	Examine theory on knowledge work as the identified formal activity of the CoP.	Examine theory on boundary conflicts to investigate relationship with wider organisation.	Examine theory on alignment and leadership to investigate effects of contributing to top-down KM.
	Plan	Plan workshop. Design interview schedule.	Design interview schedule.	Plan workshop. Design interview schedule.	Plan workshop. Design interview schedule.
Collect data	Action	Workshop.	n.a.	Workshop.	Brainstorming, workshop.
	Method	Workshop observations. Reflective interviews.	Interviews.	Workshop observations. Reflective interviews; follow-up interviews after 18 months.	Workshop observations. Reflective interviews.
Analyse		Analysed CoP internal concepts. Identified issues related to membership and trust, power-related concerns.	Analysed the task according to the TbKM framework in the context of task and subject matter, and the role of CoP in knowledge work.	Analysed a boundary conflict situated in the context of knowledge work support and knowledge flows.	Analysed workshop and interviews to identify the effects of empowerment and ownership on CoP.
Reflect	Conclusion	Confirmed: group under study meets the characteristics of a CoP.	TbKM framework applicable with some adaptations.	Conflicting thought worlds of CoP and wider organisation are multidimensional.	Alignment of informal CoP with formal strategy. Identified the role of management in top-down/bottom-up KM.
	Concepts	CF1 components: management, CoP, relationships, knowledge work.	CF2 components: refined knowledge work and added outcomes, output, and resources.	CF3 components: added 'other organisations' and relationship to CoP.	CF4 components: extended top-down KM components. Refined relationships.
	Learning	Complex relationships with various entities of the wider organisation.	The role of a CoP in knowledge work lies in peer support and sharing subject matter expertise.	Organisational dependencies influence knowledge work support and knowledge flows.	Reinvigoration of CoP through leadership and empowerment.

Source: Adapted from Koeglreiter (2009).

Fourth, if cycles are designed to be conducted independently of each other, more time can be taken to thoroughly examine the extant literature, combined with reflections on findings. This assists in gaining deeper theoretical insights over an extended period. This is difficult to achieve in AR projects that might require significant results to be achieved over a short period in order to establish organisational change momentum.

Fifth, the complexity associated with comprehensive conceptual frameworks requires that the researcher is able to adequately present the theory-building process as a coherent piece of work. Action research studies have been criticised for failing this requirement, due to their primary objective of solving an organisational problem. Structured-case with action interventions, by its highly structured nature, however, forces the action researcher to return to the relevant existing theory and clearly outline the contribution to the evolving theory throughout the data collection process.

Finally, the combination of SC and AR is supported by an extended set of research evaluation criteria (Klein and Myers, 1999; Koeglreiter 2009; Narayanaswamy and Grover, 2007).

In a project conducted as action research, each research cycle is expected to involve an intervention. For the study reported in this chapter this would have meant that significant theoretical findings extracted from an intervention-free cycle would not have been achieved. The combination of SC and AR gives the researcher flexibility to choose whether or not action is required to achieve the theoretical goal, but presents intervention-free cycles in the same format and as carrying equal significance to those cycles that include interventions.

Action research enhances SC, complementing the theoretical focus with a practical perspective. Interpretive studies based on interview data might be limited, as the researcher has to rely on the stories told by participants. Observations made in participant and participatory research—typical for AR—might assist the researcher to paint a richer picture of the organisational situation and prompt them to take into consideration subtleties that might be missed in interview situations.

Structured-case reminds the action researcher of the theoretical focus. At each research cycle relevant literature is consulted, which might draw on a variety of knowledge domains. In a pure SC approach, literature may be consulted only in the initial stages of the project, thus maintaining a narrow focus on the case throughout the process. In contrast, the changing nature of AR might require consultation of additional literature that is applicable to the problem situation of the current cycle, thus enhancing the case through taking a more systemic view.

All of this might be considered useful by IS researchers engaging in longer-term projects to assist them with documenting the theory-building process in a practice-oriented setting.

In concluding, it is noted that future research embedding action interventions within SC cycles, conducted in research contexts other than knowledge management, would assist in establishing the wider applicability of the research approach that has been described here.

References

Baskerville, R. L. (1999). Investigating information systems with action research. *Communications of the Association for Information Systems, 2*(3), 1–32.

Baskerville, R. L., & Wood-Harper, T. A. (1996). A critical perspective on action research as a method for information systems research. *Journal of Information Technology, 11*(3), 235–46.

Baskerville, R. L., & Wood-Harper, T. A. (1998). Diversity in information systems action research methods. *European Journal of Information Systems, 7*(2), 90–107.

Blum, F. (1955). Action research—a scientific approach? *Philosophy of Science, 22*(1), 1–7.

Carroll, J. M. (2000). Understanding the early stages of the requirements engineering process in practice. Doctoral thesis, Swinburne University of Technology, Melbourne, Vic.

Carroll, J. M., & Swatman, P. A. (2000). Structured-case: a methodological framework for building theory in information systems research. *European Journal of Information Systems, 9*(4), 235–42.

Eden, C., & Huxham, C. (1995). Action research for the study of organizations. In S. Clegg, C. Hardy & W. Nord (eds), *Handbook of Organization Studies* (pp. 526–42). London, UK: Sage Publications.

Gaventa, J., & Cornwall, A. (2001). Power and knowledge. In P. Reason & H. Bradbury (eds), *The SAGE Handbook of Action Research: Participative inquiry and practice* (pp. 172–89). Thousand Oaks, Calif.: Sage Publications.

Glaser, B. (1992). *Basics of Grounded Theory Analysis: Emergence vs forcing.* Mill Valley, Calif.: Sociology Press.

Jorgensen, D. L. (1989). *Participant Observation: A methodology for human studies*. Thousand Oaks, Calif.: Sage Publications.

Kemmis, S., & McTaggert, R. (2000). Participatory action research. In N. K. Denzin & Y. S. Lincoln (eds), *Handbook of Qualitative Research* (pp. 567–605). [Second edn]. Thousand Oaks, Calif.: Sage Publications.

Klein, H. K., & Myers, M. D. (1999). A set of principles for conducting and evaluating interpretive field studies in information systems. *MIS Quarterly, 23*(1), 67–93.

Koeglreiter, G. (2009). Community of practice and the wider organisation: a multi-faceted KM framework. Doctoral thesis, Deakin University, Melbourne, Vic.

McKay, J., & Marshall, P. (2001). The dual imperatives of action research. *Information Technology & People*, 14(1), 46–59.

McKay, J., & Marshall, P. (2007). Driven by two masters, serving both. In N. Kock (ed.), *Information Systems Action Research* (pp. 131–58). New York, NY: Springer.

Mumford, E. (2001). Action research: helping organizations to change. In E. M. Trauth (ed.), *Qualitative Research in IS: Issues and trends* (pp. 46–77). Hershey, Pa: Idea Group Publishing.

Narayanaswamy, R., & Grover, V. (2007). A critical assessment of information systems action research. In N. Kock (ed.), *Information Systems Action Research* (pp. 327–53). New York, NY: Springer.

Neuman, W. L. (2006). *Social Research Methods: Qualitative and quantitative approaches*. Boston, Mass.: Allyn and Bacon.

Orlikowski, W. J., & Baroudi, J. J. (1991). Studying information technology in organizations: research approaches and assumptions. *Information Systems Research, 2*(1), 1–28.

Schultze, U. (2000). A confessional account of an ethnography about knowledge work. *MIS Quarterly, 24*(1), 3–41.

Strauss, A., & Corbin, J. (1998). *Basics of Qualitative Research: Techniques and procedures for developing grounded theory*. [Second edn]. London, UK: Sage Publications.

Trauth, E. M. (1997). Achieving the research goal with qualitative methods: lessons learned along the way. In A. S. Lee, J. Liebenau & J. I. DeGross (eds), *Information Systems and Qualitative Research* (pp. 225–45). Philadelphia, Pa: Chapman & Hall.

Vidgen, R., & Braa, K. (1997). Balancing interpretation and intervention in information system research: the action case approach. In A. S. Lee, J. Liebenau & J. I. DeGross (eds), *Information Systems and Qualitative Research* (pp. 524–41). Philadelphia, Pa: Chapman & Hall.

Wenger, E., McDermott, R., & Snyder, W. M. (2002). *Cultivating Communities of Practice—A guide to managing knowledge*. Boston, Mass.: Harvard Business School Press.

Yin, R. K. (1994). *Case Study Research: Design and methods*. [Second edn]. Thousand Oaks, Calif.: Sage.

8. The Unit of Analysis in IS Theory: The case for activity

Helen Hasan
University of Wollongong

Sumayya Banna
University of Wollongong

Abstract

In the field of information systems (IS), researchers use and adapt existing theories to make sense of their data. They also build new theory from their research findings. The way theory is used, adapted or created usually assumes a certain unit of analysis, which could be the artefact, the system, the organisation, the user, the developer, the team or something else. In this chapter, we propose that 'activity' should also be considered as a suitable unit of analysis for theory in IS since the purpose of any information system is to facilitate the activities for which it is used. To support this proposition, we describe the basic tenets of activity theory and how they can be used to underpin IS research. We illustrate these with the interpretation, through activity theory, of a health information system development that aimed at identifying and meeting the needs of various users' activities. We make the claim for activity as an appropriate unit of analysis when using existing theory in IS research and when building new theory for information systems.

Introduction

In the field of IS, researchers regularly use existing theories from more established disciplines to interpret or make sense of their data. They also adapt or combine these theories to create new theoretical frameworks in order to make them more appropriate to the particular requirements of IS research. In addition, IS researchers build new theories of various types (Gregor, 2006) from their research findings.

The way theory is used, adapted or created usually assumes a certain unit of analysis, which could be the artefact, the system, the organisation, the user, the developer, the team or something else. We are not suggesting there is anything wrong with having theories that are built around these different units of analysis. Indeed, we believe that it is appropriate for a multidisciplinary field such as IS to have multiple theories addressing a range of units of analysis. We propose that 'activity' should be considered as one of the suitable units of analysis for theory in IS since the purpose of any information system is to facilitate activities. To explicate this proposition, we draw on the tenets of activity theory—an established and respected theory of human activity that has been around for nearly a century, long before the advent of computers. The foundational work of activity theory was published in Russian and translated into English only many decades later (for example, Leontjev, 1981; Vygotsky, 1978). As with any theory, it has its own concepts and language, with English words (particularly subject, object, action, activity) only approximations of their Russian counterparts.

There has already been a substantial body of work in IS and related fields that makes use of activity theory or adaptations of it. Examples include the work of Bødker (1991b), Engeström (1987), Gould (1998), Kaptelinin (1996), Korpela et al. (2000), Kuutti (1991), Star (1998) and Suratmethakul and Hasan (2004). In the next section, we present the lessons we draw from this body of work on how the tenets of activity theory can be used to underpin IS research. We describe the relevant concepts and the language of activity theory in the third section and then illustrate the use of activity theory in IS with an interpretation of a health information system development that aimed to meet the needs of various users' activities.

Finally, we draw conclusions about how our proposal might inform theory building in information systems. We consider that the activity-theoretical framework proved useful for describing a multifaceted web-based information system, its users' activities and their unmet needs, so we propose that, with activity as the unit of analysis, IS research and practice can be described in a systematic way that holistically represents purpose, dynamic context, mediation by tools and contradictions within and between activities as they interconnect.

Lessons from the Use of Activity Theory in Previous Research

Activity theory is sometimes referred to as the Russian 'general systems theory'. As is evidenced by the seminal works of Engeström (1987), Leontjev (1981) and Vygotsky (1978), activity theory is holistic, comprehensive and convincing. It has been shown to be suitable for rigorous academic studies in many fields, and IS-related research has shown that it is particularly suitable for studies of real-world practice. The word 'activity' is a translation from the Russian word '*deyatelnost*' that conveys a coherent system of human 'doing', including physical or external behaviour and internal mental processes that are combined and directed to achieve conscious goals (Bedny and Meister, 1997). According to activity theory, activities are the significant things people (the 'subjects') do and are usually long-term projects. Each activity has a purpose (the 'object') that might be concrete/real (for example, to build a technical artefact) or abstract/ideal (for example, to set up an information system). The motives of an activity are always considered to be objective, in the sense that the 'object' of an activity incorporates the motives of its 'subject', whether the activity is real or ideal (Christiansen, 1996). Activities can be carried out by an individual or a group of people who might have different motives for being involved and different understandings of what is being done. Activities might equally well be carried out by different sets of 'actions' (for example, you might entertain guests by cooking a meal at home whereas I might take them out).

An activity is the minimum meaningful context for understanding individual actions and, unless the whole activity is the unit of analysis, the analysis is incomplete (Hasan, 1999; Kuutti, 1996). Overall, this principle highlights the importance of studying human activities in context, which is of direct relevance to fields of research dealing with socio-technical systems such as human–computer interaction (HCI) and information systems. Generally, systems are designed to serve a purpose or to support user activities, so a theoretical framework is required to form the basis by placing the user and the user's activities in context, rather than placing the system itself at the centre of the evaluation process. Kuutti (1996) suggests that activity theory can provide this theoretical framework.

Crawford and Hasan (2006) contend that the main reason for the use of activity theory in IS research is that it provides a well-developed framework for analysing the complex and dynamic settings that typically involve ongoing interactions between human (subject) and technical elements (tools or objects). The theory of activity shows the effects of tools and the environment on human actions, reactions and behaviour in work settings and in users' relations with technology (Kaptelinin, 1996; Kuutti, 1996; Nardi and O'Day, 1999). 'Activity Theory, with its focus on accumulating factors that affect the subjective interpretations, the purpose, and sense making of individual and group actions and operations, provides a useful paradigm for the ways in which human experience, needs and creativity shape the design and effectiveness of emerging technologies' (Crawford and Hasan, 2006, p53).

The information technology (IT) artefacts that support information systems have evolved at a rate that makes their use particularly difficult to study. Activity theory can meet this challenge as activities are not static or rigid entities;

they are under continuous change and development (Kuutti, 1996). Historical development is not linear or structured in a predictable pattern. It is, rather, irregular and discontinuous (Kuutti, 1996). As each activity develops over time, parts of older activities remain embedded in the development process (Kuutti, 1996). Therefore, in order to understand a current activity, it is important to analyse its historical development. Activities are dynamic and in a continuous state of evolution, with development taking place at all the different levels of an activity (Kuutti, 1996). By analysing the elements, it is possible to gain an insight into this evolutionary development process and situate the activity in its historical context.

An activity is always purposeful even if the subject is not fully aware of that purpose. For example, a manager's motivations for using an executive information system might include the desire to be better informed and to make better decisions, but might also include the desire to increase status or to impress one's competitors, along with a variety of other motives (Hasan, 1998). Whether the object is material (physical) or ideal (mental) has a value in itself because it fulfils some human need (Kaptelinin, 1996). Manipulating and transforming a shared object into an outcome over a period is what motivates the very existence of a purposeful activity (Kuutti, 1996). An object only reveals itself in the process of doing and, hence, the object is continuously under development and revealed in different forms for different participants of an activity (Engeström, 1990).

Information systems projects are notoriously subject to conflicts and contradictions and activity theory anticipates this. Different individuals performing or doing an activity might have different motives for doing so, and the motives for carrying out an activity might change over time (Kaptelinin, 2002). For example, if the object of a system development project is to construct a system to make processing more efficient, the motives for doing so might vary from cost reduction (from a manager's perspective) to improving customer care (from a marketer's perspective). The concept of contradictions is core to activity theory and a key attribute of activity systems (Engeström, 1987, 2001). These can be simple conflicts, problems, historically structured tensions, virtual disturbances, gaps, dilemmas, clashes and breakdowns that provide opportunities for innovations and changes to an activity (Engeström, 2001). The absence of a well-balanced activity system, not in equilibrium because of the presence of the contradictions, can be the driving force for change (Engeström, 2001; Kuutti, 1996). In order to analyse the development of an activity system it is important to identify and resolve contradictions. If the tensions between the elements of an activity system are identified then it becomes possible to reconstruct the system in its concrete diversity and richness, for its future development (Engeström, 1999b).

An Overview of Activity Theory

Activity theory is a complex conceptual framework that has evolved historically, and continues to evolve as it is applied in research and in practice. In the 1920s, the Russian Vygotsky undertook a comprehensive study of higher mental functions and human consciousness that laid down the foundation of what is called the cultural-historical activity tradition. Vygotsky (1978) believed that the higher psychological function in humans, which is consciousness, differs from the preconscious psyche of animals and is constructed through the communication and interrelationship between subjects (people) and the objective world. Moreover, Vygotsky (1978) proposed that consciousness is not constructed through direct interactions with the world, but that the relationship between humans and objects of the environment is mediated through the use of tools (artefacts) or, in other words, that the direct association between stimulus (S) and response (R) is mediated by tools. This idea was crystallised in Vygotsky's triangular model of a 'complex mediated act' (Vygotsky, 1978), which depicts the relationship between subject, object and mediated artefact as shown in Figure 8.1.

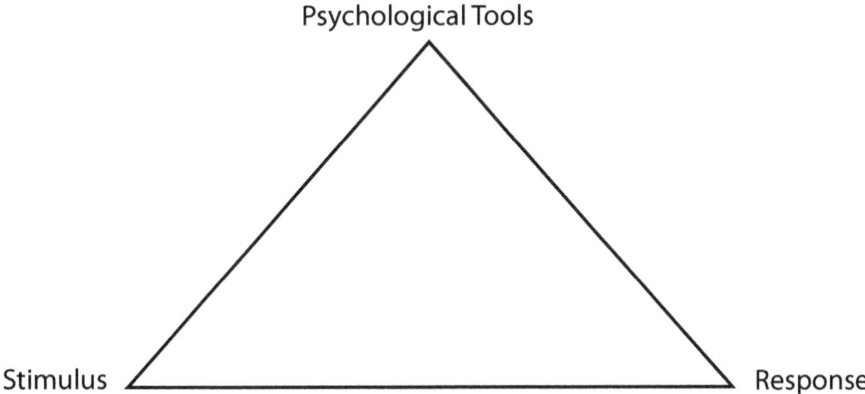

Figure 8.1 The Vygotskian Triad of Mediated Action

In activity theory, the basic unit of analysis of all human endeavours is activity, which is a broader concept than individual goal-oriented actions (Hasan, 1999). While the initial notion of an activity was generally a physical one, a later notion includes mental activities and incorporates Vygotsky's idea of mental tools as mediators, rather than being limited to only material tools of work. An activity is directed towards an object and defined by it and thus, activities are distinguished according to their object. An activity is then seen as a system in which the structure of activity is not a reaction in itself but a 'system of interrelationships' between people that is mediated by the use of instruments and tools (Verenikina and Gould, 1998). This indicates that all human activity is purposeful, is carried out through the use of 'tools' and is socially mediated. What forms the central core of an activity is the dialectic relationship between the subjects (human) and objects (purpose). After Vygotsky's untimely death in the 1930s, his colleagues, Leontjev, Luria and others, began studying human consciousness using the activity approach (Cole, 1996), and these psychologists had a profound effect on the way the theory developed. The essential principles of activity theory now include activity as the basic unit of analysis, object orientation, tool mediation, history and development, the dual concept of internalisation/externalisation, the zone of proximal development and contradictions and conflicts (Engeström, 2005; Kuutti, 1996). Most importantly, it was Leontjev who developed Vygotsky's work into a coherent, integral and conceptual framework for a complete theory of human activity (Leontjev, 1981).

Leontjev's three-level model of activity places 'activity' at the top of the hierarchy shown in Figure 8.2. An activity does not exist without a long-term purpose and strong motives whereas actions are always directed towards specific short-term goals. Participating in an activity involves performing sets of actions and operations. There might be different legitimate sets of actions and operations that will enable subjects to fulfil the purpose of the activity. An action is a conscious representation of progress towards a desired outcome, which consists of an intentional characteristic (what must be done) as well as an operational characteristic (how it can be done). According to Leontjev (1981), an operation is something that is performed routinely in order to complete an action in the current situation and condition. Operations might be performed subconsciously or automated in technology.

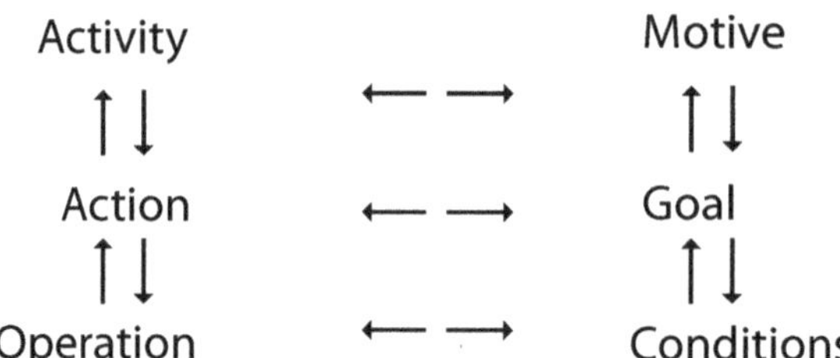

Figure 8.2 Leontjev's Three-Level Model of Activity

Engeström (1987, 1999a) proposed an enhanced model of the Vygotskian triangle, with additional elements as shown in Figure 8.3, to enable an examination of systems of activity at the macro-level of the community. This expansion of the Vygotskian triangle represents the social or collective elements in an activity system as being community, rules and division of labour. Community consists of all subjects involved in doing the same work or who work collectively. Rules mediate the relationship between subject and community and cover the conventions, regulations and social relations within the community that guide the activities and the behaviours in the system. In addition, the relationship between the community and the objects is mediated by the division of labour. This representation of activity also distinguishes between its object or purpose and its outcomes, which may be intended or unintended. In our analysis, we use this popular representation of activity.

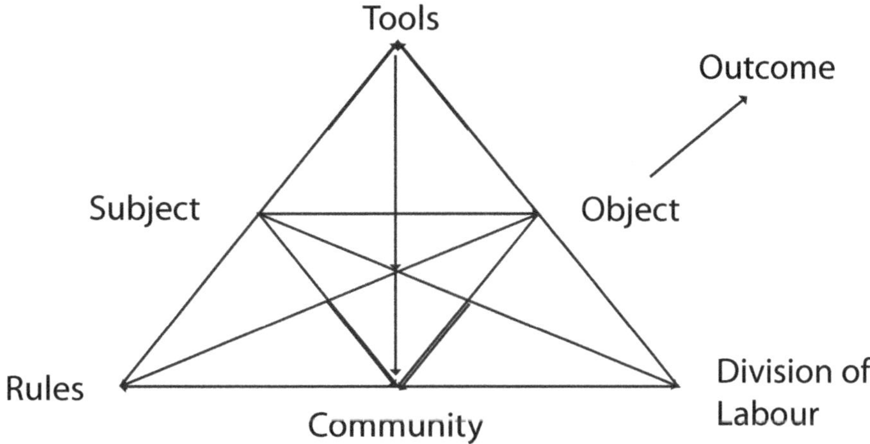

Figure 8.3 Engeström's Structure of Collective Human Activity

Source: Engeström (1999a).

Engeström (1999a) incorporates both internal and external tools in his model of activity as a system, in which internal tools would be the absorption of the inherited culture by learning and training and external tools are the new creations and inventions. An activity can have an individual as subject or can be an engagement of a collective subject composed of a group of people who would bring different skills and understandings oriented by a certain goal or common object that transforms activity into outcomes.

Cole (1999) identified as a limitation of activity theory its insensitivity towards cultural diversity and proposed that it is no longer sufficient to focus on isolated activities. When he applied the framework internationally, the question of diversity and dialogue between different cultures and traditions became a serious challenge. As a result, current users of activity theory make use of conceptual tools for analysing and transforming networks of interacting activity systems and for understanding the dialogues, multiple voices and multiple perspectives. Figure 8.4 shows one type of interaction between multiple activities—namely, where two activities have parts of their object in common (for example, the design and use of IT artefacts as described in the seminal work on participatory design of Bødker, 1991a; and Bødker and Gronboek, 1996), depicted as two interlinked activities.

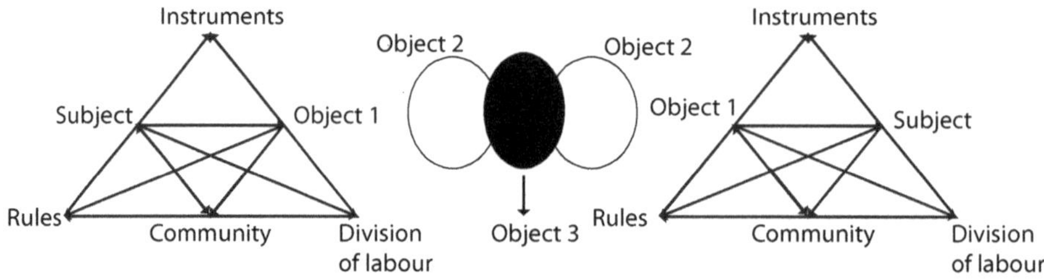

Figure 8.4 Two Interacting Activity Systems as a Minimal Model

Source: Engeström (1999a, 2001).

Engeström (2001, p. 136) described the object of an activity as 'a moving target, not reducible to conscious short term goals'. This implies that there is a demand for joint and collective work that should be established between different sets of stakeholders, governed by rules and divisions of labour, to determine the new object of interacting activity systems. Following the work of Engeström (1987) and Kuutti and Virkkunnen (1995), an analysis of an activity system normally begins with the identification of one central activity that is the focal point of holistic investigation and is surrounded by other interrelated activities that support the central activity (Hasan, 2003), as shown in Figure 8.5.

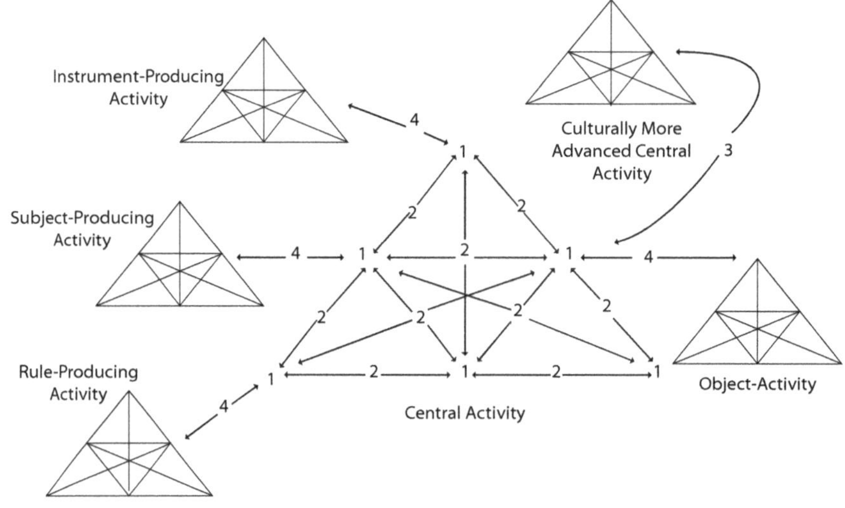

Figure 8.5 A Central Activity and Interconnected Activities

Source: Engeström (1999a).

Engeström (2005) summarised activity theory using five fundamental principles.

- The unit of analysis related to the network of other activity systems is defined in terms of its collectiveness, artefact mediation and object orientation.
- Activity systems are multi-voiced and have a community of multiple perspectives resulting from division of labour amongst the participants.
- The problems of an activity system can be understood through its development and history.
- As tensions accumulate within and between activity systems, contradictions play a central role in the change and development of those activity systems.
- Through a zone of proximal development, activity systems can transform expansively to reconceptualise the object of the activity.

Tool mediation, on which Vygotsky based his original work, is often considered the most fundamental principle of activity theory. It is the use of tools that distinguishes human activity from the activities of animals. Leontjev (1981) asserts that tool mediates activity and thus connects humans not only with the world of objects but also with other people. Because of this humans' activity assimilates the experience of humankind.

An activity is mediated by different types of tools: the tools used and the social context of the work activity. The two-way concept of mediation implies that the capability and availability of tools mediate what can be done and the tool, in turn, evolves to hold the historical knowledge of how a society works and is organised (Hasan, 1999). Human activity is mediated by a number of tools (external and internal). Tools specify modes of operation and are historically developed in social terms, possessing an evolutionary cultural component. An activity is defined by the tool-mediated relationship between subject and object—that is, between the doer and their purpose. Tools expand our potential to manipulate and transform objects, but also restrict what can be done because of the limitations of the tool, which, in turn, often stimulates improvements to the tool. The mediation is a mutual development of both the activity and the kinds of tools used.

There are three kinds of tools that mediate human activity (Bertelsen, 2000; Hasan, 1999; Hasan and Gould, 2001)

- primary tools (artefacts, instruments, machines, computers, and so on)
- secondary tools (language, signs, models, ideas, and so on)
- tertiary tools (cultural systems, scientific fiction, context, virtual realties, and so on).

Since primary tools are physical (material tools), they produce changes to the material object, whereas the secondary tools (psychological tools) influence the psyche and behaviour of subjects. Regardless of the type, however, all tools are transmitters of cultural knowledge (Kaptelinin, 1996) or a historical residue of activity development (Kuutti, 1996). Tools determine the modes of operation and are historically developed possessing cultural aspects. As such, the use of these culture-specific tools shapes the way people act (Hasan, 1998; Nardi, 1996). In this sense, this aspect can shape future designs of systems. When the tools are computer based, this notion becomes a source of power (Kaptelinen, 1996), especially when used in the context of analysing the dialectic interactions between people and technologies, and how they are shaped by human activity.

Application of Activity Theory to the Study

The authors were recently involved in research into the design and implementation of a web-based health information system (hereinafter referred to as the 'Health IS') to support the provision of health information to the medical community and the public. This e-health study will be used here to illustrate the power of activity as a unit of analysis. As mentioned above, activities of design and the use of IT artefacts have been the object of study in HCI and information systems. In this mode of research, the basic activity model is expanded to include a minimum of two interacting activity systems, as shown in Figure 8.4. The design activity is constrained by the computer in various ways, through the actually available materials as well as through the past experiences of designers and users (Bødker et al., 1987). Designers must have primary data about real activities that various users engage in rather than relying solely on their own prior knowledge and experience, and the system functions, to define user tasks. The Health IS can be depicted as the outcome of a 'technical design' activity and a 'tool for the use' activity. The use of participatory design methods where end users are invited to participate in the development of the IS system is currently widespread in the healthcare sector (Pilemalm and Timpka, 2008).

In order to make a better design and ultimately to create better health IT-based artefacts, designers and users undertake a number of interrelated and somewhat overlapping activities that in our case also involved the researchers. The experiences, resources, tools, and so on of designers meet, and sometimes clash, with those of the users, and with others involved. In our concern for the web of activities involving a particular IT-based artefact, the design activities are essential and should emphasise how our understanding needs to reach beyond the immediate use (Bødker, 1991b).

In the course of the project, the researchers created many diagrams to visualise the interconnections between the activities they were observing. An example is depicted in Figure 8.6, which shows the partial overlap between the objects of two activities in the manner of Figure 8.4. The bottom triangle is the *design* activity and the top triangle the *use* activity of the new Health Information System. A common motive of both activities is to improve healthcare outcomes through shared IS tools. There were, however, some differences in the tools, communities of practice and intended outcomes, with the *design* activity more concerned with efficiencies and reduced costs through the use of the Health Information System.

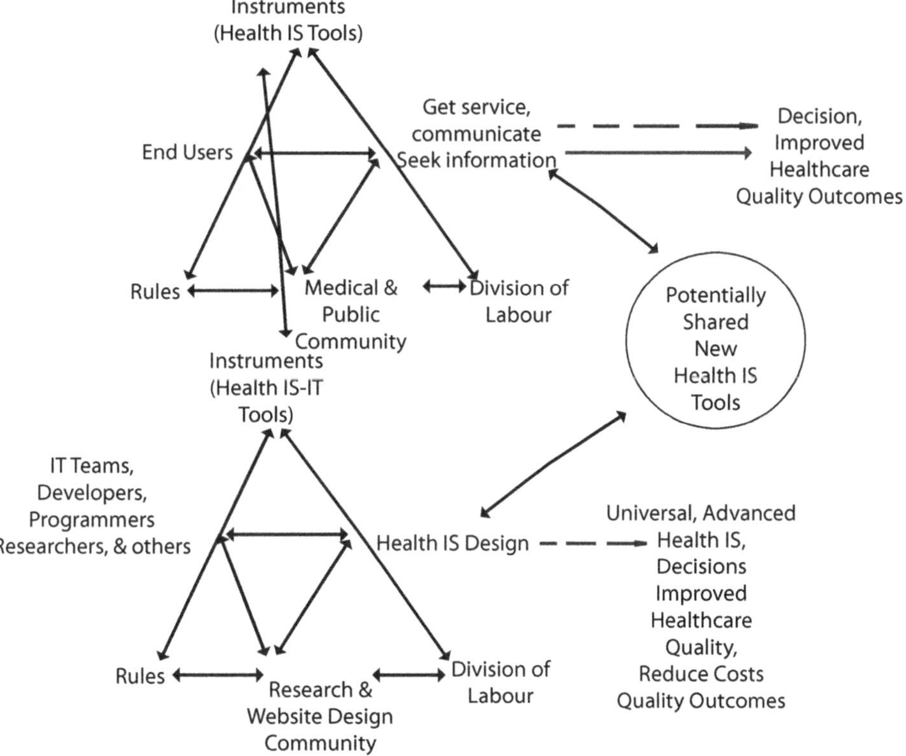

Figure 8.6 One Interpretation of the Interacting Activities of Design and Use

Developing the Research Activity

We saw our study as depicted in Figure 8.6 but were initially restricted to a study of the use activity only. We therefore began the process of applying the activity theory framework by mapping out 'use' as the central activity and then moving to the surrounding interrelated activities, one of which was the design activity where our findings on the use activity would help the web site designers.

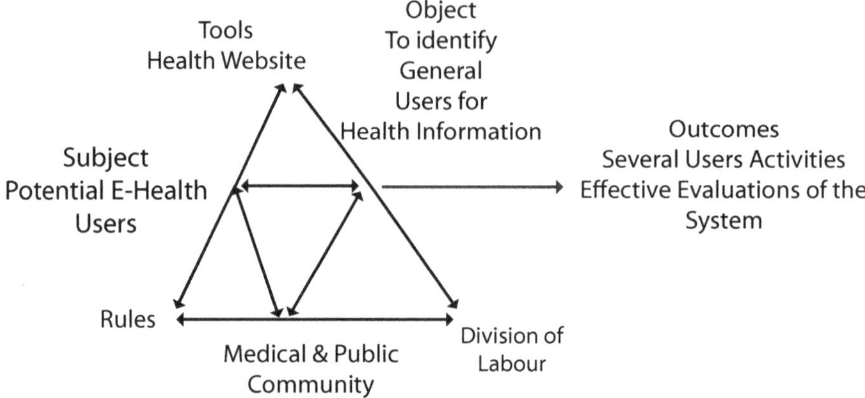

Figure 8.7 The Activity of General Users of Health Information

Our initial depiction of the activity of end users, for which the Health IS would be a tool, is shown in Figure 8.7 and it was proposed to conduct usability tests on the current web site as a form of action research. Usability testing typically involves carefully producing scenarios to reflect realistic situations in which a person carries out the required tasks using the system being evaluated and tested while the observer or the researcher watches and takes notes. These tests soon revealed that there could be several different activities of use and therefore different activity systems based around these.

The research itself was considered an activity of the research team that was connected to the activities being studied through participatory action research. According to Engeström and Kerosuo (2007), an interventionist researcher must find dialogue partners who share their emotions, concerns and agendas within the activity system. For this reason, we turned to Q Methodology as a discovery-mediating tool since it allows the researcher to open up and dig into the subjective views of the participants in a study. It places the participants at the centre of analysis and enables the researcher to explore ways to engage and motivate people. Figure 8.8 depicts the framework for the research activity as used for this study.

A detailed description of this research and the results of the Q analysis have been published elsewhere (Banna et al., 2010). It is sufficient for our purpose here to report that a three-factor solution was considered the best candidate for interpretation of the data. We labelled these factors, in order from most highly to most lowly ranked, as: 1) 'service-oriented users'; 2) 'interactive users'; and 3) 'information seekers'. These factors were then reinterpreted as activities since, to make sense of the results of the Q study, we took each factor as the unit of analysis and reinterpreted them using the concepts and language of activity theory.

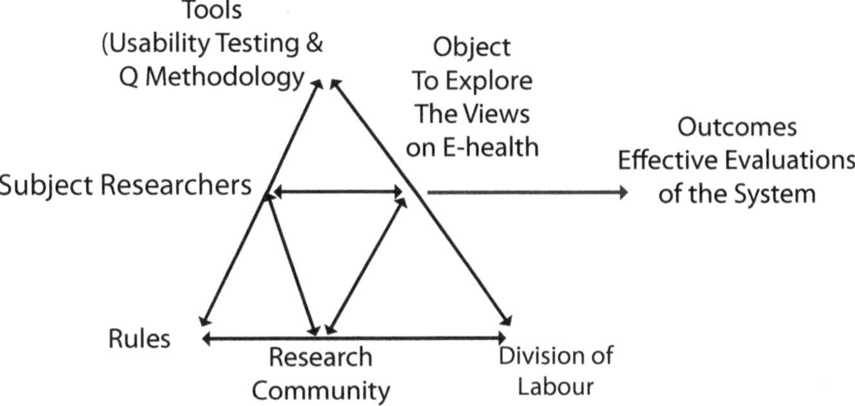

Figure 8.8 The Research Activity

The Activities of the Users

Drawing on our previous experience with Q methodology, we assumed that each of the three sets of users participates in a different 'use' activity. As subjects of that activity, they have distinct characteristics, have a particular object in mind when they use the Health IS and therefore use a different version of the tool. In other words, the Health IS web site would need to be designed differently in each case. We now describe the activity that each group of users would carry out when they used the Health Information System.

Figure 8.9 shows the 'service-oriented users' in an activity that is bound to the object of getting health services-related information. Those who were located on this factor in the Q study were mainly tertiary students with different majors and degrees (many doing medicine) and medical staff. As the subjects of this activity, they are intelligent and knowledgeable in medical and health service matters. Obtaining specific health-related services information for themselves or others is the object that defines the activity. Making better healthcare decisions is the most common outcome of this activity. Their activity is mediated by the community, which includes well-educated people internal and external to the healthcare system, but knowledgeable of it.

The activity of the 'interactive users' is shown in Figure 8.10. These subjects are not passive recipients but active participants. Their active use of the Internet focuses on a desire to engage in communication. They wanted to use the Health IS to interact with experts, to mutually determine what might be best for themselves as well as others, to engage in collective decision making with regard to tasks and to access information and resources. The outcomes of this interactive activity should allow users to create, share and manage knowledge, skill sets and attitudes needed to cope with the dynamic nature of healthcare settings and circumstances. These people included local and international academics, healthcare workers, palliative care staff and the general public, although it is interesting to note that healthcare workers and palliative care staff made up almost 50 per cent of this group. It seems that healthcare workers naturally want to work in teams.

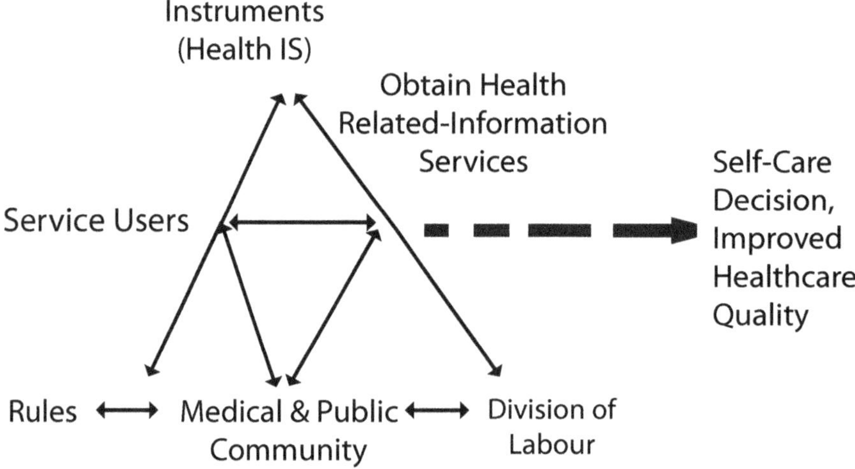

Figure 8.9 The Activity of Getting Information Related to Health Services

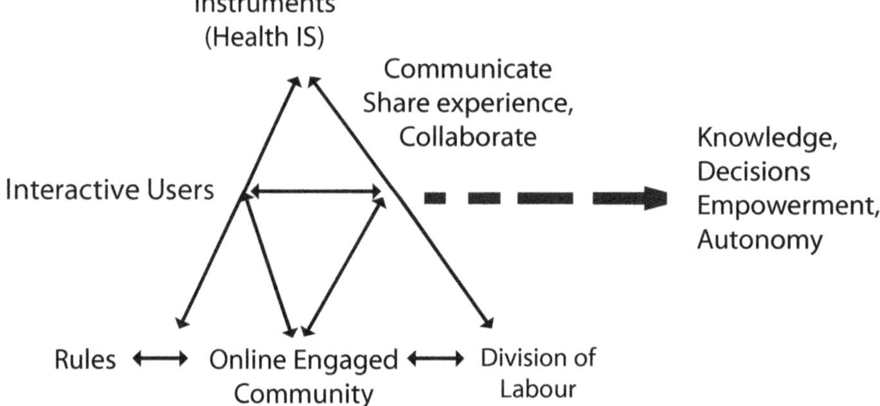

Figure 8.10 The Interactive Communication Activity

Figure 8.11 illustrates the activity of the 'information seekers'. Their core activity is to explore the Internet by themselves to find information. They see health web sites as information-intensive portals that should target a variety of users and enable them to make better health choices and decisions on their own. The subjects of this activity were mainly local and international students and a mix of university staff members.

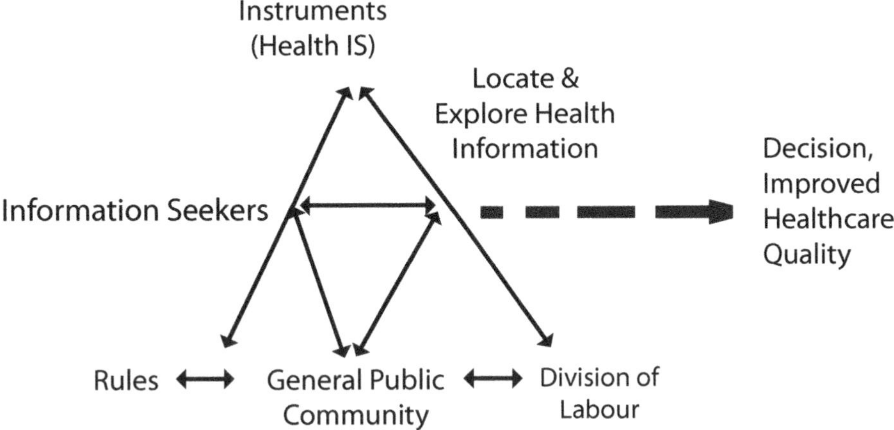

Figure 8.11 The Activity of Seeking Health Information

Once we had reinterpreted the Q study factors as activities, we could then apply other concepts of activity theory to gain a deeper understanding of the phenomena under investigation. This process is illustrated in the next section of the chapter.

Application of Activity Theory Principles

Identifying the Mediating Tools for the Activities

The way tools mediate activities is a key activity theory principle and one particularly significant in IS research and practice. There needs to be consideration of the primary, secondary and tertiary tools that mediate the various activities that are carried out, in this case by the diverse users of the Health Information System. As explained earlier, primary tools are physical and tangible whereas secondary and tertiary tools are psychological and cultural, reflecting and influencing the behaviours of the subjects (Hasan and Gould, 2001). While a web site has obvious physical characteristics—namely, the content and the way it is structured—it is also a secondary tool for the user. The information, knowledge and sense-making it provides should be viewed as tertiary tools. The quality of the physical tool can determine the quality of psychological tools, which are necessary for generating social interaction through a rich representation of information and making communication more effective among healthcare workers on the one hand and between health experts and the public on the other.

Primary tools. A web site acts as a location of primary tools for all user activities: seeking health information, interaction, communication and the exchange of information, and enables users to obtain health-related services. The primary tools of interest for information seekers are general search engines. The primary tools among service-oriented users are also search engines and perhaps other more specific navigation tools. In contrast, interactive users are more likely to use social technologies such as email, discussion forums, chat rooms, weblogs and online community services. The term 'Web 2.0' reflects the ongoing transition of the World Wide Web from a simple collection of web sites to a fully fledged computing platform serving these social web applications to end users. Their low cost and connectivity functionality are the lures of these social technologies. They also support new forms of informal network interaction and activity between people, enabling and enhancing informal access to ways of creating and disseminating information.

Secondary tools. These include medical and health knowledge, communication skills, previous or past experiences and specialised language, which might be different for each group of subjects. Healthcare staff use their own medical and professional language, while general users use common non-medical language. Public health web sites also have to consider the multicultural identities and backgrounds of users and so provide information in different languages. Language issues and ways of presenting information can make the design activity more difficult as developers of the system need to design for all levels of language skills. Information brokers might be needed to help in the design. For interactive users, the web site can incorporate new social technologies that enable ordinary people to have a global presence, giving users a new flexibility and independence to support collective actions, knowledge sharing and decision making by self-directed groups.

Tertiary tools. These include, most importantly, the social context. In health-related matters, the context is often stressed as users wanting to find and communicate health information concerning their own medical condition or that of a loved one. Stress reduces cognitive capacity and this must be considered in designing the web site, particularly for the service-oriented users. As noted above, healthcare providers constitute more than 50 per cent of the interactive user group. It might be important for healthcare providers to create virtual communities to disseminate the required health information and circulate their ideas and knowledge among themselves. This could result in better decision making and knowledge management that in turn improves healthcare outcomes.

Different activities and different types of tools soon make an activity system diagram quite complicated. For example, if we revisit the simple design-use activity system of Figure 8.6, we might start to add other activities as shown in Figure 8.12. Here the Health IS is depicted as the outcome of both a technical health IS design activity, which considers it as a primary tool, and the data collection activity performed by the information brokers, which considers the web site as a secondary tool. A link between the objects of the Health IS design and data collection activities represents the communication and cooperation that is needed if the Health IS is to be both technically sound and provide the right kind of information. In Figure 8.12 a feedback loop has been added from outcomes of the use activity to the link between the design activity and the data collection activity. This feedback loop is particularly important to ensure that the goals of multiple voices or multiple perspectives are met.

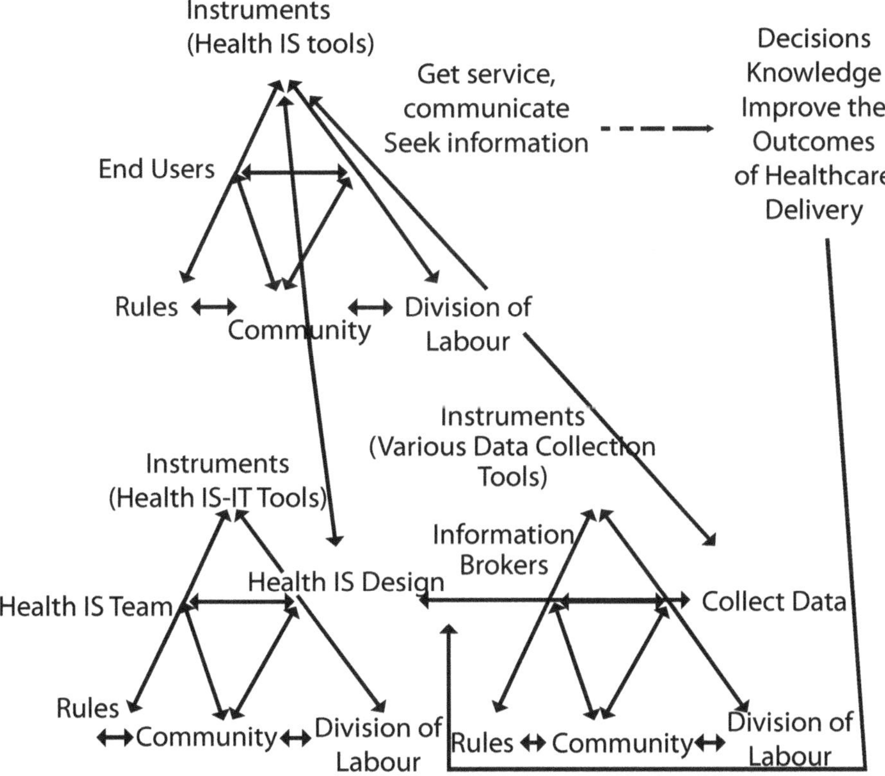

Figure 8.12 The Interrelated Activities of Data Collection, the Use and the Model

Internalisation and Externalisation

Activities have a dual nature because they have an internal and an external side (Kuutti, 1996). When external mediating tools, such as health web sites, are integrated into functional organ and goal-oriented configurations, they are perceived as an attribute of the individual, implying that they naturally extend the individual's abilities, thus shaping the boundary between internal tool (based inside the human mind) and external tool (the outer world). It is this that distinguishes between expert users and novice users of the Health Information System. The merging of internal and external tools is evident in expert users who use the Health IS as a seamless extension of their abilities.

In novice users, who are still learning how to use the Health IS, the boundary between the internal and the external tools is the most apparent. The boundary between the internal (human mind) and the external worlds becomes less clear and distinguished when users repetitively use tools to carry out an activity. In other words, expert users are deemed to have internalised more of the central activity as well as the Health IS itself. Users of the Health IS make decisions based on information from external sources, including primary care providers, health web sites, and so on, all of which is internalised and manipulated in the subject's internal plan of action using mental models or maps. This is a dynamic situation and an understanding of this can shape the future design of the system. It creates a challenge for developers to design a system to meet the facts of multiple voices and multi-perspectives of users (novice and expert). Designers rarely consider how the tools they design will mediate activities, and change work practices and social and cultural norms (Hasan, 1999). Mediating tools modify and transform the learner's thinking processes as they begin to use new tools to express their thinking (Cole and Wertsch, 2001). When there is internalisation of an external activity and mastery of the existing Health IS, users tend to develop a need for new mediating Health IS tools. This is so because an activity system is unstable and dynamic, making the design activity an ongoing process.

The Principles of Contradiction and Conflict

In activity theory, contradiction and conflict are seen as the sources of learning and development. Because of the dynamic nature of activity systems, the Health IS must be designed for change. Therefore, the development of Health IS projects must include processes for user participation and feedback and implementation of new requirements. Within an activity system, there are different people with different backgrounds, motives and perspectives. The notion of multiple voices, described earlier, can be a source of conflict in the design activity but it can also lead to positive action if there is a forum for voicing different user views. It is important to expose multiple voices to negotiation and change when understanding and improving an activity system.

Despite the potential for contradictions to change and transform the activity system, this transformation does not always happen. In fact, contradictions can either enable change or disable change. This depends on whether they are identified, acknowledged and resolved. Hidden, invisible or un-discussable contradictions are the most difficult to identify and these tend to be taken for granted among design teams. From this perspective, to enable innovations, the resolution of contradictions cannot happen at the individual level; it requires social interactions. Human–computer interaction researchers such as Bødker (1991b) have recognised that there has to be close collaboration and cooperation between the use activity and the design activity. This implies that these are in a continuous cycle of change where computer applications as well as other parts of work activities are constantly reconstructed using different design tools. A clear knowledge of the changes paves the way to better design.

When analysing tensions, Engeström (1987) proposed four levels of contradiction. Level-one contradictions are breakdowns within and between the elements of action that make up the activity and are affected by other related activities. This means that the same action can be executed by different people for different reasons or by the same person conducting two separate activities.

Secondary inner contradictions are those that occur when users of the system encounter a new element of an activity. The process for incorporating the new element into the activity brings conflicts. For example, designers might face difficulties assimilating and coordinating user requirements and new rules of government and the division of labour. Tertiary contradictions occur between the existing form of an activity and what can be described as a more advanced form of the activity. This might occur when the design activity is reconstructed to take account of new motives, new tools, new user skills or new ways of working. Quaternary inner contradictions are tensions between the central activity and related activities like, for instance, instrument-producing, subject-producing and rule-producing activities linked to the central activity of the system.

Research in the field of IS takes into consideration the dynamic interplay between information and communication technologies, activity and uses, and patterns of human experiences that emerge over time as the dimensions of the whole system of work activity changes (Crawford and Hasan, 2006).

Conclusion

We have proposed activity as a suitable unit of analysis for theory-based research in IS, where activity is understood in terms of the concepts and language of activity theory. A recent study by the authors has been reinterpreted as a system of activities to illustrate the value of applying an activity-based framework to IS settings. We applied activity theory because of its holistic and contextual emphasis that is appropriate for qualitative and interpretative research exploring how organisations understand and meet the challenges of designing IS artefacts. In particular, activity theory is known as a well-developed framework and a powerful tool for analysing and providing deep and rich understandings of complex and dynamic settings such as occur in the public healthcare context. This approach relies on taking activity as a holistic and complex unit of analysis, offering a unique lens through which to analyse behaviours, processes, tools and outcomes.

In several of our studies, the combination of activity theory and Q Methodology has proven an appropriate technique for conducting IS research and interpreting results in an integrated holistic way. The factors that come out of the Q study invariably relate to specific activities of the people who hold similar views on a topic. In this case, examining those activities with the rich concepts of activity theory contributed to an overall understanding of users' perceptions and the purposes of their different activities of use of the Health Information System. Indeed, this leads to the more general observation that humans use diverse information systems on a daily basis to achieve their personal and/or work objectives, with the expectation that these information systems will facilitate the activities in which they are engaged. Therefore, activities represent a basic element of the context in which systems must exist and operate. Indeed, we contend that attempting to understand information systems is pointless without also attempting to understand the activities in which they are involved. Information systems only become meaningful in the context of use and, in order to successfully undertake the design activity, the use activity must be taken into account.

Activity theory can be used in its traditional form or adapted in ways not anticipated by its founders. For example, the triangular representation of an activity originated by Engeström (1999a) is a relatively recent adaptation but has formed the basis of many studies into complex organisational settings. The concept of an activity, however, remains as a dialectic relationship between subject and object (someone doing something) mediated by tools of various kinds. We suggest that activity as a unit of analysis could be the basis of new theory. In general, we note that in using, adapting or building theory consideration should be given to the unit of analysis it assumes.

References

Banna, S., Hasan, H., & Meloche, J. (2010). Subjective evaluation of attitudes towards e-health. *Proceedings of AIM Conference*, Kuala Lumpur, Malaysia.

Bedny, G., & Meister, D. (1997). *The Russian Theory of Activity: Current applications to design and learning*. Newark, NJ: Lawrence Erlbaum Associates.

Bertelsen, O. W. (2000). Design artefacts: towards a design-oriented epistemology. *Scandinavian Journal of Information Systems, 12*, 15–27.

Bødker, S. (1991a). Activity theory as a challenge to system design. In H.-E. Nissen, H. K. Klein & R. Hirscheim (eds), *Information System Research: Contemporary approaches and emergent traditions* (pp. 551–64). Amsterdam, Netherlands: Elsevier.

Bødker, S. (1991b). *Through the Interface: A human activity approach to user interface design*. Hillsdale, NJ: Erlbaum.

Bødker, S., & Gronboek, J. (1996). Users and designers in mutual activity: an analysis of cooperative activities in systems design. In Y. Engeström & D. Middleton (eds), *Cognition and Communication at Work* (pp. 130–58). Cambridge, UK: Cambridge University Press.

Bødker, S., Ehn, P., Kammersgaard, J., Kyng, M., & Sundblad, Y. (1987). A utopian experience. In G. Bjerknes, P. Ehn & M. Kyng (eds), *Computers and Democracy—A Scandinavian Challenge* (pp. 251–78). Aldershot, UK: Avebury.

Christiansen, E. (1996). Tamed by a rose: computers as tools in human activity. In B. Nardi (ed.), *Context and Consciousness: Activity and human–computer interaction* (pp. 175–98). Cambridge, Mass.: MIT Press.

Cole, M. (1996). Cultural psychology: a once and future discipline. In V. de Keyser (ed.), *Work Analysis in French Language Ergonomics: Origin and current research trends. Volume 34* (pp. 653–69, 1991). Cambridge, Mass.: Harvard University Press.

Cole, M. (1999). Cultural psychology: some general principles and a concrete example. In R. Miettinen, Y. Engeström & R. L. Punamäki (eds), *Perspectives on Activity Theory* (pp. 87–106). Cambridge, UK: Cambridge University Press.

Cole, M., & Wertsch, J. V. (2001). *Beyond the Individual-Social Antimony in Discussions of Piaget and Vygotsky*. Retrieved from <http://webpages.charter.net/schmolze1/vygotsky/colewertsch.html>

Crawford, K., & Hasan, H. (2006). Demonstrations of the activity theory framework for research in IS. *Australasian Journal of Information Systems, 13*(2), 49–68.

Engeström, Y. (1987). *Learning by Expanding: An activity-theoretical approach to developmental research*. Helsinki, Finland: Orienta-Konsultit.

Engeström, Y. (1990). When is a tool? Multiple meanings of artifacts in human activity. In Y. Engeström (ed.), *Learning, Working and Imagining: Twelve studies in activity theory* (pp. 171–95). Helsinki, Finland: Orienta-Konsultit.

Engeström, Y. (1999a). Expansive visibilization of work: an activity-theoretical perspective. *Computer Supported Cooperative Work, 8*(1–2), 63–93.

Engeström, Y. (1999b). Innovative learning in work teams: analyzing cycles of knowledge creation in practice. In R. Miettinen, Y. Engeström & R. L. Punamäki (eds), *Perspectives on Activity Theory* (pp. 377–404). Cambridge, UK: Cambridge University Press.

Engeström, Y. (2001). Expansive learning at work: toward an activity theoretical reconceptualization. *Journal of Education and Work, 14*(1), 133–56.

Engeström, Y. (2005). Developmental work research: expanding activity theory in practice. *International Cultural-Historical Human Sciences. Volume 12.* Berlin, Germany: Lehmanns Media-LOB.

Engeström, Y., & Kerosuo, H. (2007). From workplace learning to inter-organizational learning and back: the contribution of activity theory. *Journal of Workplace Learning, 19*(6), 336–42.

Gould, E. (1998). Psychological information systems frameworks: a contrast between cognitive science and activity theory. In H. Hasan, E. Gould & P. Hyland (eds), *Information Systems and Activity Theory: Tools in context* (pp. 39–58). Wollongong, NSW: University of Wollongong Press.

Gregor, S. (2006). The nature of theory in information systems. *MIS Quarterly, 3*(30), 611–42.

Hasan, H. (1998). Activity theory: a basis for contextual study of information systems in organizations. In H. Hasan, E. Gould & P. Hyland (eds), *Information Systems and Activity Theory: Tools in context* (pp. 19–38). Wollongong, NSW: University of Wollongong Press.

Hasan, H. (1999). Integrating IS and HCI using activity theory as a philosophical and theoretical basis. *Australian Journal of Information Systems, 6*(2), 44–55.

Hasan, H. (2003). Communities as activity systems and other such frameworks. In H. Hasan, I. Verenikina & E. Gould (eds), *Information Systems and Activity Theory. Volume. 3. Expanding the horizon* (pp. 74–95). Wollongong, NSW: University of Wollongong Press.

Hasan, H., & Gould, E. (2001). Support for sense making activity of managers. *Decisions Support Systems, 31*(1), 71–86.

Kaptelinin, V. (1996). Activity theory: implications for human–computer interaction. In B. Nardi (ed.), *Context and Consciousness: Activity theory and human–computer interaction* (pp. 103–16). Cambridge, Mass.: MIT Press.

Kaptelinin, V. (2002). Making use of social thinking: the challenge of bridging activity systems. In Y. Dittrich, C. Floyd & R. Klischewski (eds), *Social Thinking: Software practice* (pp. 45–68). Cambridge, Mass.: MIT Press.

Korpela, M., Soriyan, H. A. & Olufokunbi, K. C. (2000). Activity analysis as a method for information systems development: general introduction and experiments from Nigeria and Finland. *Scandinavian Journal of Information Systems, 12*(1), 191–210.

Kuutti, K. (1991). Activity theory and its applications to information systems research and development. In H.-E. Nissen, H. K. Klein & R. Hirscheim (eds), *Information Systems Research: Contemporary approaches and emergent traditions* (pp. 529–49). Amsterdam, Netherlands: Elsevier.

Kuutti, K. (1996). Activity theory as a potential framework for human–computer interaction research. In B. Nardi (ed.), *Context and Consciousness: Activity theory and human–computer interaction* (pp. 17–44). Cambridge, Mass.: MIT Press.

Kuutti, K., & Virkkunen, J. (1995). Organisational memory and learning network organisation: the case of Finnish labour protection inspectors. *Proceedings of the Twenty-Eighth Hawaii International Conference on System Sciences (HICSS), Wailea, HI, 4*, 313–22.

Leontjev, A. N. (1981). *Problems of the Development of the Mind*. Moscow, Russia: Progress Publishers.

Nardi, B. (1996). *Context and Consciousness: Activity theory and human–computer interaction*. Cambridge, Mass.: MIT Press.

Nardi, B., & O'Day, V. (1999). *Information Ecologies: Using technology with heart*. Cambridge, Mass.: MIT Press.

Pilemalm, S., & Timpka, T. (2008). Third generation participatory design in health informatics—making user participation applicable to large-scale information system projects. *Journal of Biomedical Informatics, 41*(2), 327–39.

Star, S. L. (1998). Working together: symbolic interactionism, activity theory and information systems. In Y. Engeström & D. Middleton (eds), *Cognition and Communication at Work* (pp. 296–318). Cambridge, UK: Cambridge University Press.

Suratmethakul, W., & Hasan, H. (2004). An activity theory analysis of a case of IT-driven organisational change. *Proceedings of the IFIP TC8/WG8.3 International Conference: Decision Support in an Uncertain and Complex World.*

Verenikina, I., & Gould, E. (1998). Cultural-historical psychology and activity theory. In H. Hasan, E. Gould & P. Hyland (eds), *Information Systems and Activity Theory: Tools in context* (pp. 1–18). Wollongong, NSW: Wollongong University Press.

Vygotsky, L. S. (1978). *Mind in Society*. Cambridge, Mass.: Harvard University Press.

9. IT-Driven Modernisation in Agriculture: New theories for a new phenomenon[1]

Lapo Mola
University of Verona

Cecilia Rossignoli
University of Verona

Walter Fernandez
The Australian National University

Andrea Carugati
University of Aarhus

Abstract

The information systems (IS) literature has largely neglected the study of implementations of large-scale strategic initiatives to modernise the agricultural business. This chapter reports an ongoing empirical study of the efforts of a multibillion-dollar organisation to modernise the operations of its supplier base. Modernisation, as an external force affecting organisations, is a new and different phenomenon with respect to organisational change that is normally considered as resulting from internal effort. Modernisation is an ongoing, evolving process performed by organisations in order to survive and prosper. Yet the decision to modernise is likely to face the forces of entrenched traditions and practices: the feelings and social significance of established ways of those with the power to derail the modernisation project. Using institutional theory as the theoretical lens through which to study the role of information and communication technologies (ICT) in modernisation strategy, this chapter argues that managers should take care to go beyond the reasons for change, considering as well the physical, social and cultural needs of the stakeholders involved. Our study extends the literature on agribusiness management by highlighting the tensions between the initiator of a modernisation effort and the suppliers who will need to adapt and respond to it.

---

1 A previous version of this chapter was presented as a paper at the Academy of Management Meeting, Montreal, Canada, 6–10 August 2010.

Current Issues and Trends in IS and Agribusiness

The impact of technology on an organisation is a classic topic in information systems (IS) research. Typically, however, the level of analysis is the organisation itself, or teams or individual people within it. The tendency is to move to 'lower' levels of analysis with ever finer research instruments. Consequently, IS theories often fail us when we want to theorise, not about an organisation wanting to implement a particular new internal system, but rather about an organisation that wants to change an entire sector or industry. This problem is not uncommon, with examples ranging from a government that wants to change the way its employees work (Senyucel, 2008) or that might want to implement the same information technology (IT) system in all of its municipalities (Kaylor et al., 2001), to a supermarket chain wanting all of its suppliers to use the same supply chain system (Stalk et al., 1992) and a manufacturer demanding that all of its suppliers pass through the same Internet portal (Klein and Krcmar, 2006). The challenges of these cases are different because the 'victims' of the implementation initiative are not under the direct control of the initiative taker, and because they have their own specific way of working, and might even have specific cultures or traditions that might align against the initiative. This process is, thus, not that of classic organisational change but instead what sociologists call *modernisation*.

In the past this problem was dealt with mainly as a strategic problem related to the bargaining power of buyers and suppliers. When electronic data interchange (EDI) was the leading technology in supply chain modernisation, a classic warning was issued: 'EDI your suppliers before your client EDIs you.' Modernisation through technology was a strategic matter. Despite the fact that this still holds true, it is only one side of the coin. Information technology-driven modernisation processes have for the most part failed (Klein and Krcmar, 2006; Rossignoli et al. 2009), mostly because, behind the bargaining power, their transformative implications were not well understood. Our interest lies in taking seriously these large-scale transformative processes to try to understand their adoption and diffusion dynamics. Of particular concern to us is the continuous process of adaptation in which the modernising organisation uses ICT as a significant and transformative tool to modernise a sector or industry populated by organisations that will be on the receiving end: the 'modernisees'. Rather than taking a functionalist focus, we are interested in how the stakeholders interact during a modernisation process and how the result of this interaction leads to a new state desired by the moderniser, or derails the modernisation attempt. Our guiding research questions are:

- How do people in the modernising organisation enact modernisation strategies?
- How do people in the modernisee organisations react to modernisation strategies?
- How do stakeholders' actions contribute to (or detract from) the modernisation initiative?

To start the process of finding answers to these questions, we adopt an interpretative epistemology using mainly qualitative data from case studies (Walsham, 1995). Adopting a particular research paradigm also implies the adoption of a system of beliefs about the ontological, epistemological and methodological nature of the world and how to investigate it (Guba and Lincoln, 1994). We use institutional theory (Scott, 2008) as a theoretical lens through which to make sense of our findings since our data analysis showed that modernisation replicates many of the traits of institutional change.

Our chapter aims to contribute to academic and practical knowledge of modernisation processes in agribusiness, which is an industry that is critical to an ever-growing population with an increasing demand for food both in terms of quality and quantity. It is puzzling to us that, while we depend on the success of agribusiness to cover the most basic of our human needs, the agribusiness sector has largely failed to attract the attention of information systems and management researchers. The lack of academic interest seems particularly prevalent in developed economies where only a handful of papers report on modernisation of business practices in agribusiness (for example, Gregor and Jones, 1999; Lindsey et al., 1990). Further, studies involving innovation are often conducted from a perspective of technical rationality that is not always appropriate to non-technical processes in which problems and solutions are unclear, confusing and conflicting (Schön, 1983). In our particular case, the modernisation process offers many uncertainties and the role of ICT is ill defined. Thus, we aim to engage in a discourse that considers the existence of conflicts between ends and means. We agree with Schön (1983, p. 127) that '[i]t is rather through the non-technical process of framing the problematic situation that we may organise and clarify both [the] ends to be achieved and the possible means of achieving them'.

The rest of the chapter is organised as follows. We first present a brief discussion of the literature dealing with modernisation practices and the role of ICT in those practices, including the modernising of traditional industries such as agribusiness. Then, we present institutional theory as a lens through which to make sense of modernisation and, before presenting in detail the case under study, discuss the nature of the research project, its delimitation and the methodological approach followed. The discussion focuses on the tension between conflicting demands and the rationale for a favourable outcome. We conclude by arguing for the importance of this type of study, note its limitations and describe the nature of the next phase of research.

Modernisation Practices

While modernisation has largely escaped the attention of IS researchers (perhaps because of the organisation or user focus of IS research), its meaning of 'being modern' is fertile ground for the work of sociologists and philosophers. In sociology, modernisation is conceived as a process of social evolution that brings societies inexorably from simple coordination and control, production and distribution structures to ever-greater levels of development and civilisation, usually characterised by increasing complexity. This evolution has been perceived as a Darwinian process of survival of the fittest (Spencer and Carneiro, 1974; Turner, 2000), where modernisation implies a change from a present state to a future state, possibly *progress* towards a better state according to some measure. In societal terms, modernisation aimed at achieving a 'better' state is culturally defined, and not typically universally accepted. Acceptance of the 'better' state depends on the observer: the moderniser or the modernisee. Its nature might be viewed from a functional perspective that is related to the achievement of a more productive state (Spencer and Carneiro, 1974) or it might be social, focusing on more emancipation or more equal opportunities and where modernisation is 'a progressive force promising to liberate humankind from ignorance and irrationality' (Rosenau, 1992, p. 5).

The terms *modernity* and *modernisation* are of course tightly connected; modernity defines the state and modernisation the process to reach that state. Connected as they are, these terms are still intensely debated by sociologists. But while the ontological debate about the meaning, nature and purpose of both modernity and modernisation is outside the scope and focus of this chapter, we nevertheless need to define our views. By modernising, we imply a departure from past practices, not just a change. We also accept that modernisation is 'a state of consciousness which defines the present in its relation to the past' (Delanty, 2006, p. 82).

While the sociological debate focuses on mega-trends and speaks of unavoidable consequences and unstoppable floods, management theories are focused on control and performance, so, not surprisingly, discussions about modernity are largely absent from IS research. Nonetheless, the connection between technology and 'the way things get done' has been discussed both by academics (Malone and Laubacher, 1998; Scott Morton, 1971) and by lay observers (Friedman, 2006).

The connection between ICT and modernity should not be surprising. Over time, we have grown accustomed to the belief that ICT plays a central role in modernising organisations and societies alike. Yet Misa (2004) points out that, albeit with a few exceptions, this belief is not supported by sufficient theoretical work, and only lately have researchers begun to theorise about the entanglement between the concepts of ICT and modernisation.

ICT and Modernisation

The implementation of ICT in organisations might be explained from two alternative perspectives: the functionalist and the subjective (Markus and Robey, 1988). The functionalist perspective sees the process of modernisation as an inevitable process of adaptation to technological, societal and structural changes, where the dominating decision factor is the achievement of improvement in performance—of some form or other. Modernisation is therefore seen as a rational process of pursuing ever-increasing levels of performance. In the functionalist view, performance and its increments can be measured and therefore progress is inevitable and non-compliance is seen as resistance or sabotage and it is therefore sanctionable (Bain and Taylor, 2000; Foucault, 1977; Sewell, 1998). As societies evolve, they experience more and more problems of coordination, production of goods, services and knowledge, and finally in finding ways of distributing these resources. Latour (1993) defines the modernisation process as one of purification and translation. As complexity in society increases, the need to divide elements emerges—humans from non-humans—and to interconnect them again. In functionalist terms, the effort is therefore one of optimisation of subsystems and the coordination among them.

The subjectivist view sees the modernisation process as a less predictable path led by the multiple interpretations of reality that individuals develop. Modernisation is not seen as unidirectional increments, but change is instead seen as an organic process where individuals continuously adapt to emerging interpretations of their world across multiple dimensions (Cecez-Kecmanovic et al., 1999; Zuboff, 1988). Taking a subjectivist perspective, the results of introducing technology into organisations depend more on the interpretation of the technology by the actors involved than on the intrinsic characteristics of the technology itself (Orlikowski, 1993). In this view, whether *progress* is reliant on technological innovation depends on the observer and is open for debate.

Despite criticisms and different perspectives, technology and its development have traditionally been linked to the process of modernisation. Not surprisingly, it has been argued that the eras of human modernity have coincided with new discoveries that have radically changed the structure of society (Misa, 2004). For example, the Industrial Revolution had at its core technological innovations such as the steam engine, textile machinery and production techniques, and also the use and sharing of technological knowledge (Senyucel, 2008). The combination of these elements provided the seeds of the contemporary interpretation of modernity, including the hierarchical structuring of organisations and the division of labour.

While the early periods of IS research had an overly functionalist view of the role of IT in organisations, the later period, which arguably started with Markus (1983) dealing with power and politics in Management Information System (MIS) implementations, is marked by a more balanced view. Today we observe that IS research on the role of IT in modernisation is divided into three groups: those who propose that technological evolution will lead organisations through certain modernisation paths (Gibson and Nolan, 1974; Scott Morton, 1971); those who propose that individual agents will shape the technology to their own will (Cecez-Kecmanovic et al., 1999; Zuboff, 1988); and those who posit that individuals will use technology among many other tools in their practices, and ultimately practices will shape structures or be shaped by them (Barley, 1986; DeSanctis and Poole, 1994; Orlikowski, 2000).

At present, there is no theory of modernity that embraces these three views, especially when the locus of interest is not just the organisation but also an entire industry. Furthermore, while we know that dedicated modernisation efforts can influence the way people work, there is still a need for theorising on the connections between ICT and efforts to change deeply entire industries that have so far managed to remain impervious to the otherwise pervasive flood of information and communication technologies.

ICT and the Modernisation of Agribusiness

If studies of ICT as a tool of modernisation are scarce, those looking at ICT as a tool for modernisation in agribusiness industries are even more so. Among the very few we were able to find, and perhaps the closest to our field of interest, is an action research study conducted by Gregor and Jones (1999) that reports on the successful development and adoption of electronic commerce in the Australian beef industry. They use the theory of diffusion of innovations (Rogers, 1995) as their theoretical lens and contribute to our understanding of communication and communication channels in terms of adoption.

A contrasting case, also in the Australian beef industry, is offered by a study of Business to Business (B2B) marketplaces that reports a low rate of adoption. Factors contributing to the poor adoption rate included the difference of perceived advantages at group and at individual levels, the users' reluctance to change traditional ways of trading, the power of the users to opt (or not) for the new system and, finally, the slow pace of adoption creating a situation in which a critical mass of users was not achieved in a timely manner (Driedonks et al., 2003). Rogers' theory of the diffusion of innovations was again used, although this time in conjunction with Kambil and van Heck's (1998) model of exchange processes. The study found that Kambil and van Heck's model does not suit 'situations in which the adoption decision is optional' (Driedonks et al., 2003, p. 37) and this finding is relevant to our study because we are also dealing with powerful (or at least not powerless) potential adopters, as the description of our study and foundation case will explain.

Institutional Pressures and Modernisation

While Gregor and Jones (1999) used diffusion theory, they were not exactly focusing on modernisation processes but rather on the diffusion of e-commerce in the beef industry. Modernisation has a diffusion component but, as explained above, it is a very multifaceted phenomenon. It is characterised by a central organisation wanting to change, with technology, the behaviour of many supplier organisations that are typically very entrenched in their practices and very conservative users of ICT as a management tool. This is a story of traditions spanning organisational boundaries, similar behaviours influenced by laws, norms and networks, and a seemingly unjustified resistance to technology. Institutional theory provides us with the instruments to make sense of these instances and to understand technology-driven change (Barley, 1986). Multilevel analysis (Scott, 2008) can help us understand the influence of society and the network on the individual firm, isomorphic forces can help to understand horizontal changes (Di Maggio and Powell, 1983) and rationalised myths can help to understand institutionally driven interpretations.

The main aspect that characterises agribusiness is the cohesiveness among members and the need for coordination to move perishable products quickly and in a controlled way through the supply chain. These stringent needs—like the necessity to continuously control the cold chain for some products—require not only coordination but also trust along the supply chain. This trust is often enacted through professional networks or, as in our case, in very large cooperative agreements regulated by contracts and rules.

A way to regard this industry is to consider its institutional elements. Trust, in particular, can be explained according to the culture and traditions of the organisations involved. By investigating this issue through an institutional theory lens, we might better understand the persistent influences on how institutions emerge, survive, change and dissolve. This approach also notes the pervasive influence of institutions on human behaviour, including processes by which rules, routines, norms and belief systems guide social behaviour (Scott, 2008; Svejvig and Carugati, 2010).

Social actions exist in accordance with or in reaction to complex, durable, resilient social structures (Currie, 2009), governed by organised, established procedures and rules that constitute the essence of institutions. In modernisation attempts, we can therefore expect that institutions and the rules of the game will affect what the modernisees will and can do. Institutional rules can guide actions in certain directions independently of or in complete contrast with the modernisation effort (Vitharana and Dharwadkar, 2007).

Institutional theory highlights the mechanism of isomorphism—'a constraining process that forces one unit in a population to resemble other units that face the same set of environmental conditions' (Di Maggio and Powell, 1983, p. 72). Therefore, similar organisations in the same environment tend to pursue similar courses of action. Isomorphism derives from three underlying pressures in the institutional environment: mimetic, normative and coercive.

Mimetic pressure arises from similar interpretations of the organisational field, such that organisations mimic other successful organisations they consider similar to them (Di Maggio and Powell, 1983). The assumption is that what has worked in the past for one organisation will continue to work for other organisations. The decision to implement a specific information system or a specific business process often depends more on the environment—that is, on what other similar organisations are doing—than on any objective needs that are parts of the modernisation processes.

Normative pressures induce isomorphism through the shared respect for unwritten codes of conduct or traditions. Professions and industries absorb such traditions, often by hiring people from similar educational backgrounds. Over time, normative pressures permeate the business and dictate what should be done and how to approach problems. For example, when a firm realises that most of its competitors are adopting a specific protocol and procedure, it will tend to adopt similar protocols and procedures to achieve legitimacy among its customers and business partners (Vitharana and Dharwadkar, 2007).

Coercive pressures instead emerge from legislation and technological changes that compel the organisation to adapt (Tushman and Anderson, 1986). For instance, the European Union introduced the requirement for milk traceability and other norms that force all the actors operating in the milk supply chain to declare all information regarding their procedures. A collective respect for these norms will therefore create equal behaviour or, in other words, isomorphism.

Isomorphism is not, however, either force or resistance. Rather, the isomorphic pressures can act both as a brake and as an accelerator of change, depending on the particular situation that a market is in and the momentum that a particular initiative is having. In our case, we note that all three pressures can be present and jointly affect the modernisation practices.

Furthermore, isomorphic pressures act at multiple levels. Scott (2008) identifies four levels of mutual influence: society, organisational field (defined as 'organisations that, in the aggregate, constitute a recognised area of institutional life' (Di Maggio and Powell, 1983, p. 1983)), organisation and individual actor. Currie (2009) encourages IS researchers to conduct multilevel analyses to enrich understanding. Organisational fields can develop professional codes of conduct, perhaps regardless of legislation, though Scott (2008) finds reciprocal interactions across levels such that the societal level connects to the individual level through the organisational field level, and vice versa. Through institutional change and the diffusion of practices, top-down processes allow higher-level structures to shape the structure and action of lower levels—something that would be desirable in modernisation processes but that can also work in the opposite direction and thwart the initiative.

Institutional theory, with its ability to highlight both change and resistance and top-down and bottom-up influences, is therefore a powerful tool for understanding the complex modernisation phenomenon.

Research Methodology for the Empirical Study

In this chapter we report the first phase (or foundation case) of a long-term, theory-making, exploratory research project. It focuses on an important agribusiness organisation that is undertaking a process of modernisation of their (arguably) conservative practices. Given the ultimate goal of theory development, we needed to adopt a methodological approach that allows for the rigorous treatment of data and for the evolution and integration of knowledge. This section briefly describes our methodological approach and the limits for the study.

Given the potential scope of the research project, it was necessary to define from the outset the limits of the study (Creswell, 1994). Accordingly, our study focuses on cases where top management decides to modernise their operations by implementing new technologies to ensure the continuing prosperity and survival of their organisation. We further limited our study to situations in which the people required to adopt new technologies and practices are: a) relatively free to adopt or to reject the new technology, and b) experience a certain cultural attachment to their *old ways* of operating. In this sense, we are seeking technologies that, while rationally necessary for the organisation to adopt, have the potential to be socially or culturally disruptive to the end users. That is, we focus on cases in which the new technology replaces an entrenched tradition, or way of doing business, among the actors. The final delimitation of the study relates to the nature of the national economies in which the organisations reside. While it is common for research on modernity to be concerned with developing countries, we wanted to observe modernisation practices in developed nations. Thus, suitable organisations for this study must be located in countries that are among the top 30 economies in terms of gross domestic product (GDP), as defined by either the International Monetary Fund (IMF) or the World Bank (WB).

To start the study, we required a suitable research site in which to conduct an exploration that would allow us to find guiding themes and relevant concepts critical to further explorations, using more relevant and sharper questions (Eisenhardt, 1989). In other words, we adopted a theoretical sampling strategy based on relevance and emergence (Glaser, 1978) rather than the random sampling strategy suitable for other approaches. The unit of analysis for our study is the modernisation attempt as a whole and this implies that we also need to analyse the technological and contextual components and actors involved in the case. We thus need both rich empirical data and access to multiple stakeholders—two critical elements for theoretical triangulation and conceptual development.

Given the importance of context to our study, we also needed to find specific cases of modernisation from which we could derive knowledge through the rigorous interpretation of actions, accounts and patterns found in the data. To achieve this we adopted an interpretative case study approach (Walsham, 1995), collecting and analysing data following the principles of classic grounded theory methodology (Glaser, 1978, 1998; Glaser and Strauss, 1967).

The approach we adopted has been described as an effective and appropriate way of researching emerging phenomena in their own organisational and human context (Orlikowski, 1993; Van de Ven and Poole, 1989). Furthermore, this approach allows us to explore the substantive area of study in order to produce grounded explanations of the phenomena under observation (as suggested by Eisenhardt, 1989; Orlikowski, 1993) that is informed but not hijacked by the extant literature and institutional theory.

The founding element for our research is a longitudinal case study that takes place in a large cooperative organisation in the economically important Italian dairy sector. From this case we have collected rich data via interviews, web site analysis, on-site observations and documents such as web site logs and electronic web services (for example, descriptions of processes and practices, news and newsletters). The next section presents our foundation case.

Case Description: Quattro and 4HQ

The foundation case follows a modernisation initiative called Quattro (a pseudonym) implemented by an association of cooperatives in the agricultural industry operating in Italy, the seventh country in the world in terms of (GDP), according to lists produced in 2008 by both the IMF and the World Bank. Quattro is a business organisation owned by four agricultural cooperatives (each owning a 25 per cent share) and has at its core the development, implementation and management of an information system we call 4HQ, where HQ stands for 'high quality'.

We selected this case based on the characteristics of the organisation with regard to the need for modernisation, the complexity of the business, the coexistence of multiple views on modernisation and the use of ICT as an enabler of modernisation. We have been able to obtain rich data and access to documents and key people in the organisation (CEO, general managers, directors, CIO and farmers).

To explain the Quattro case, the following subsections: 1) describe the organisational characteristics and its historical background, and 2) explain the driver for modernisation and the modernisation project.

The Social Enterprise of Cooperatives

According to the *Macquarie Dictionary*, a cooperative society is a 'business undertaking owned and controlled by its members, and formed to provide them with work or with goods at advantageous prices (a consumers' cooperative is owned by its clients, a producers' cooperative is owned by the workers)'. The concept of cooperatives developed in the nineteenth century. It was connected to the socialist movement and opposition to the system of manufacturing that resulted from the Industrial Revolution. The principal characteristics of a cooperative are that the participants are moved by egalitarianism and solidarity motives more than by the need to achieve personal success (Thompson, 1824). As a social structure, a cooperative can be seen as a clan, a group that shares values and beliefs, thus allowing minimisation of goal incongruence and toleration of high levels of ambiguity in performance evaluation (Ouchi, 1980).

Nowadays, cooperatives are social enterprises that take different forms and sizes and enact different systems of control and ownership (Ridley-Duff, 2009). Furthermore, cooperatives can be an important economic contributor. In Italy, cooperatives generate between 5 per cent (source Legacoop) and 8 per cent (source UNCI) of national GDP and involve about one million people across the country, touching the lives of practically every Italian. The Italian cooperatives involved in our study (the Coops) are formed and owned by more than 2000 farmers (dairy producers) and have a combined annual turnover of more than €1 billion (see Table 9.1).

Table 9.1 Participating Cooperatives' Membership Size and Annual Turnover

Cooperatives	No. of members	Turnover (€ million)	Turnover (€ million) per member—avg.
Coop 1	231	120	0.52
Coop 2	180	40	0.22
Coop 3	1600	990	0.62
Coop 4	250	18	0.07
Total	2261	1168	0.56

Members (farmers) contribute to the enterprise by contributing work or capital to the cooperative. In compensation, they receive shares and dividends from the cooperative as well as being able to access its services and resources. These farmers have two main ways of influencing decisions via the exercise of voting rights. First, they elect the members of the board of directors, who must be members of the cooperative, and the CEO (also a member) in charge of managing the cooperative. Second, acting as shareholders, farmers are able to vote during annual general meetings. It should, however, be noted that each member has the right to only a single vote, regardless of the share equity they have in the cooperative (this is part of the egalitarian doctrine of cooperatives).

The Business Environment: Market and regulatory pressures

The processing, manufacturing and distribution of dairy products across Italy and the world form the core business of the Coops. The business association of the Coops in the Quattro venture was part of a strategy to modernise their operations in response to external pressures from the European Union and from major corporate customers.

Following a new regulation introduced by the European Community (Regulation EC No. 178/2002 atr.3), which established strict rules regarding the traceability of agricultural products, the Coops felt that the food industry was under pressure to: 1) respond to consumer demands for products of high quality and reliability; 2) provide greater transparency in order to increase consumer confidence in the Italian agricultural system; and 3) introduce strict protocols and controls in the industry.

The Coops also needed to increase the efficiency of their supply chain in response to demands from corporate customers known as *Grande Distribuzione*

Organizata (*GDO*) or 'large distribution business'. The *GDO* customers are the most important distribution chains for food products (among other products) and include large distribution businesses such as COOP Italia, Esselunga, Supermerceti PAM, Conad and SIGMA. Examples of *GDO*s outside Italy include Carrefour and Auchan in France, Tesco and Marks and Spenser in the United Kingdom and Wal-Mart in the United States. Given their significant bargaining power, the *GDO* customers are often able to define the terms and conditions of transactions and they require a highly efficient supply chain.

Paving the Way: Intentions and outcomes

To address the pressures, the Coops decided to join forces in order to implement an information system that would allow a more effective integration of different actors along the supply chain. The aim of this project was to look beyond achieving higher efficiency. For the Coops, more effective integration meant better coordination and therefore a better chance to work on the quality of the products of the entire supply chain. Information and communication technologies (ICT) were perceived as the vehicle for enabling standardisation in breeding and cultivation, and thus achieving formalisation of best practices within strict protocols. Finally, the documentation and traceability of all phases in the production process were accepted by management as the natural path to achieve the level of quality required from both the modern individual consumer and the *Grande Distribuzione Organizata*.

The objective of the new system (ICT, people and processes) was therefore to combine and distribute the information necessary to ensure not only the efficiency of the system but also the high quality of the products produced by the Coops. This was to be achieved by controlling the processes of standardisation, formalisation and traceability, as shown in Figure 9.1.

The Coops represent more than 2000 farmers across Italy. These farmers are also special shareholders of the cooperatives; they participate in committees and in management positions as well as casting votes. In short, the farmer-shareholders are the owners of the cooperatives that own Quattro and also the key users (and clients) of Quattro's services.

Accordingly, the mission of Quattro is to provide services to its members that contribute to improving the quality of their work and the quality of their produce. To do that, Quattro, via 4HQ, offers services that can be categorised into two macro-areas: standard services for the supply chain and customised services.

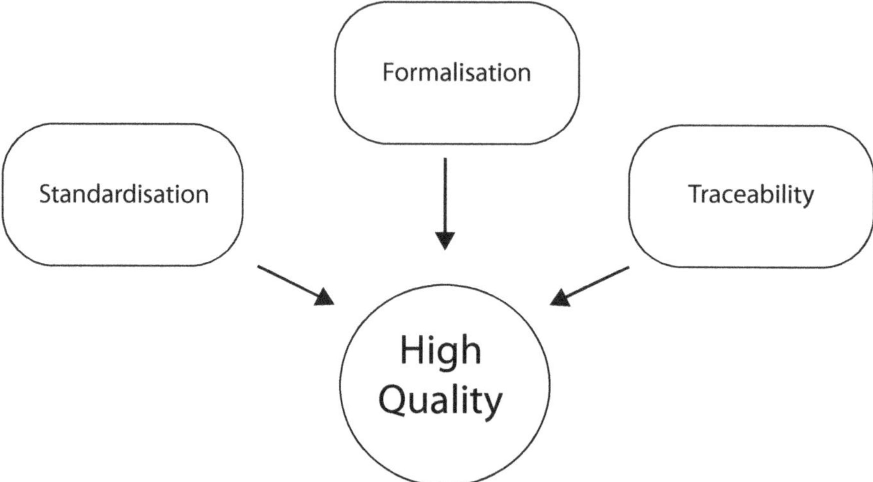

Figure 9.1 Processes Contributing to High Quality at Quattro

Informational Services

The purpose of the first type of service is to support the commercial activities of the participants through an easy, effective and consistent set of communication channels. These services target the information needs along the supply chain. The main tool is a portal that every week publishes relevant information such as new agricultural laws and regulations, protocols, farming products, and so on.

Customised Services

This second group of services consists of the ability to customise services according to the user accessing the 4HQ system. Depending on their identity and access rights, users are able to visit the portal to obtain sector-specific, technical marketing information and links to local businesses. In this category, we find the services such as

- zootecnic consultancy
- traceability information systems
- consultancy in designing and implementing management and control systems
- training and e-learning services
- research project design, coordination and implementation.

Analysis and Discussion of Intuitionalism and Modernisation in Agribusiness

The initial goal of Quattro was to use ICT to increase the number of transactions in the supply chain, as the Managing Director of Quattro explained at the time:

> In this way, it is possible to provide the companies of the sector with outsourceable support, thus creating a vertical integrated value chain and therefore allowing immediate product traceability and other benefits. The main goal in our mind was to create the new Amazon.com of the dairy industry.

With such a vague goal in mind it is not surprising to see that 4HQ and Quattro evolved not into an 'Amazon.com of the dairy industry' but into something else, by drifting, changing and adapting in a process of discovery and evolution. Nowadays, 4HQ enables a strong integration of the supply chain, from suppliers of raw materials (milk and agricultural products) to food-processing companies. The system connects, informs and provides services to more than 2000 participating enterprises—a network of mainly small and medium enterprises in the agricultural and food industries operating across the Italian territory.

The services most appreciated are those related to traceability. A considerable amount of effort and resources is focused on guaranteeing traceability and quality of the products in the supply chain. Quattro also focuses on developing a technological environment to provide timely and accessible consultancy services to participants, and thus on developing an integrated information system among farmers, cooperatives and clients that adds value to the participant parties and generates revenue for Quattro. The need for traceability is a powerful coercive isomorphic pressure that acts in support of the modernisation process.

The changes introduced by the system are not just technical; the system is changing the way farmers have been operating for a very long time. The introduced changes are forcing farmers to change, and some are more receptive than others to the new ways. One of them stated that those farmers who want to compete in today's business environment 'cannot sell their products any more on the basis of trust and their well-known name'. The new system is replacing a way of living and a source of pride—the pride of being associated with and being known for the quality of what they produce. Here there are coercive pressures supporting the modernisation process while normative pressures linked to trust and fame tend to slow the process.

The globalised market, with its demand for quality assurance, brought along new actors such as the 'certificators' EUROCAP or GLOBALGAP that control the processes of production in order to confirm certification. Some farmers think that the computers will bring an aura of professionalism and therefore will facilitate the certification.

> These certificators feel more confident if they see that we [the farmers] are using a computer for managing the information about pesticides, fertilisers or phytodrugs and demonstrate that the production process has followed the right production protocol.

The rationalised myth of computers bringing professionalism in their wake is quite widespread and certainly has some leverage. This isomorphic pressure can work in support of the modernisation process.

The second objective of Quattro was to improve the efficiency of the farmers in response to the demands of the *Grande Distribuzione Organizata*. This aim was not easy to achieve due to the small average size of Italian farms, which makes achieving economies of scale difficult. According to the Italian Institute of Statistics (ISTAT), in 2005 there were 1 728 532 farms in Italy, with an average of 10.4 ha of land, of which 7.4 ha were used for agriculture.

Resolving the Tension Between Members' Approval and Market Demands

One of the roles of the cooperatives is to provide services to their members and yet, due to the cooperative nature of organisation, this is not a typical client–service-provider relationship. The people working in and managing the Coops and Quattro have to take into account, in their service provisioning to member-owners, that their actions and decisions must ultimately be supported by the majority of the members. As a vice-president of one of the cooperatives said:

> On one side we don't have independent autonomy. We are representatives of our members; we [the cooperative] were born to serve them. On the other side, we must cope with the rules of the market otherwise the risk of bankruptcy is more than a hypothesis.

In contrast with a traditional corporation, a cooperative has to find equilibrium between two competing demands. Cooperatives act as 'normal' organisations in that they compete on the basis of efficiency and efficacy; however, a couple of extra considerations are very important. First, for production cooperatives (as in our case), their suppliers are also their owners and they therefore cannot pursue maximisation of efficiency and efficacy on the supply side without taking into account the particular social responsibility embedded in the cooperative movement, such as issues of mutual help, solidarity, social values and equality that are so important to their being.

Second, every significant decision, operational or strategic, has to be ultimately shared with and approved by members. The cooperative's board of directors is directly responsible to the members who elected them to the board. In addition, for a cooperative, the rule of 'one head, one vote' means that the strategies and the policies adopted have to be accepted by the majority of the members and reaching consensus is culturally accepted as the correct way to implement change.

Thus, *consensus building* is an important management element to consider, in part because collaboration within the boundaries of the cooperative is a natural expectation and also because initiatives aimed at improving the collaboration capability of the cooperative are (in principle) well accepted. Since consensus from the members is needed to run the cooperatives successfully, the cooperatives must endeavour to control the information flows in order to increase transparency on the condition of the market and to provide relevant and timely information to their members.

When all of this is taken into account, it seems that members should readily adopt a system such as 4HQ since, in theory at least, nobody would reject the services offered by it and there is a broad consensus on its potential value. Despite this fact, we observed that the pace of adoption of 4HQ proceeded slowly among the members.

According to people involved with technical aspects of the system, there are a few factors influencing the slow uptake of the technology, and these factors seem to be mainly related to the characteristics of the users. The 4HQ IT Manager said that, in this environment, ICT systems 'are often too complex to be useful. The lack of high-speed connection, the time and the knowledge required to use the system make the farmer uncomfortable with it.' The chief technical officer of one of the cooperatives responsible for developing part of the system points to the age and lack of computing skills of the farmers: 'On one hand there is the average age of the agricultural entrepreneurs. Most of our members belong to that generation which never used a PC or IT in general.' These claims are partially contradicted by other evidence (ISTAT) showing that, while the phenomenon of the ageing population of farmers persists, younger farmers are in control of the majority of businesses with higher income. Therefore, the supposed lack of ability with computers can easily be seen as a rationalised myth and its disconfirmation shows that the modernisation process could be easier than expected.

When we consider these facts it seems that we are observing a younger and more academically qualified farmer emerging in Italy. Given the 'one head, one vote' rule of the cooperatives, however, the greater number of small farms, perhaps with less computer literate leaders, have considerable voting power within the cooperatives. Furthermore, while younger farmers seem to be more receptive to the technology and able to persist with the system in order to get the benefits it offers, for the system to be successful wider adoption is required. We seemed to touch again on the barrier of age during an interview with a thirty-four-year-old farmer who said:

> I started using the system three years ago. It took a little time to use it properly, but eventually I was able to sort it out. Now I can use it easily. I use the system once a week. It takes from 40 minutes to one hour to fill out all the forms and do the job. It was a little bit tricky, especially at the beginning, and I understand if some of my colleagues, especially the older ones, find it difficult.

This situation is particularly worrisome at a time in which the product of agribusiness is not just the material produce, as one of the vice-presidents lamented:

> They [some farmers] think that a bottle of milk is a bottle of milk. Today this assumption is not true any more. A bottle of milk is made of a bottle, milk and all the information related to the production, distribution and stocking processes adopted and printed on the label.

How the Use of the 4HQ Can Provide Great Advantages to Cooperatives and Consumers

Achieving the goal of high quality is perceived as critical to the future of the cooperatives, both in terms of complying with laws and regulations and in terms of satisfying the demands of corporate customers such as *Grande Distribuzione Organizata*. One of the initiatives is to have an electronic log, an electronic repository called '*quaderno di campagna*' (loosely translated as 'book of practice'), which contains all the information about the processes conducted in the field. A senior manager explains both the nature of and the need for this service:

> We are the interface between the farmers and the *GDO*. If the law requires, for example, a maximum of 3 mg per kilo of a certain substance (pesticides), the *GDO* will often require a lower percentage. The amount of residuals will depend on when and how the farmers use these substances. They have to declare everything on the *quaderno di campagna*, which can be filled offline [using paper-based forms] or online. If they use the offline version it is possible that they might make a mistake—for example, writing the wrong name for the product or declaring a quantity which is inaccurate. Once a season, we collect and check all documents. This means that we can handle the problem only

> when it is too late. With the online version, we see in real time what the farmer is doing. He cannot mistake the name of products, due to the prompts the system provides. If he enters something incorrectly, we can contact him immediately and organise a proper response.

The previous quotation encapsulates both the need for and the complexity of a system in which choices (that is, doing the work offline or not at all) can jeopardise the cooperatives' attempts to reform their operations. The management of the cooperatives clearly believe that, as the agricultural industry faces an increasing level of pressure to obtain high quality of product, service and information, they must modernise or perish, and that this goal can be largely attained through a process that calls for the following:

1. *Standardisation of production procedures.* The farmers must standardise their procedures in order to be compliant with the law and protocols established by institutions, markets and governments.
2. *Formalisation of the distribution phase.* The process of distribution should be conducted in a manner in which verification and traceability are possible. This formalisation will fulfil the final client informational needs, presenting important consumer information in an accurate, relevant, standardised and specific manner.
3. *Coordination as integration among the actors.* Because of the high level of interdependency among all phases of the supply chain, the actors must be efficiently and effectively coordinated. The overall quality of the industry depends upon ensuring reliability and proper controls at each step of the supply chain. The easiest way to obtain this is to trace every step of the supply chain through sharing, validating and integrating all the necessary information.

Finally, the implementation of the 4HQ platform requires not just a good technical solution but also a significant modernisation process that takes into account the needs and culture of the farmers and not just the cooperatives' rationale for change.

Conclusion

Unlike a traditional corporation, a cooperative must deal with maintaining the equilibrium between respect for their long-established institutional social contract with the members and the efficacy and efficiency required by the market. Our chapter has described one significant initiative that addresses the need of an economically important group of cooperatives to respond to market and regulator demands for traceability, quality assurance and efficient practices.

The reported phase of the study focused on the rational reasons for the initiative, as perceived by the management of the agricultural cooperatives, in an under-researched area of management and information systems. The study highlights tensions between the very nature of the cooperative enterprise and their need to adapt and respond to pressures from a fast-changing environment. Our main contribution lies in understanding both the external pressures and the organisational response through a modernisation attempt—an evolutionary change that would, if successful, allow the organisation to prosper.

We have, so far, come to only partial conclusions. Institutional theory has been useful to highlight certain contrasting elements but we acknowledge that our analysis is just at the beginning. There is a need to further investigate how farmers respond to changes in practices that in some cases will replace longstanding institutions. Therefore, our next study will endeavour to understand the reasons for delayed adoption, focusing on understanding the behaviour and concerns of the non-adopters (who are under-represented in the current study) and how these behaviours and concerns can be addressed in the context of the agribusiness cooperatives.

Bibliography

Argyris, C., & Schön, D. A. (1996). *Organizational Learning II: Theory, method and practice*. Reading, Mass.: Addison-Wesley.

Bain, P., & Taylor, P. (2000). Entrapped by the 'electronic panopticon'? Worker resistance in the call centre. *New Technology, Work and Employment, 15*(1), 2–18.

Barley, S. R. (1986). Technology as an occasion for structuring: evidence from observations of CT scanners and social order of radiology departments. *Administrative Science Quarterly, 31*(1), 78–108.

Cecez-Kecmanovic, D., Moodie, D., Busuttil, A., & Plesman, F. (1999). Organisational change mediated by e-mail and intranet: an ethnographic study. *Information Technology & People, 12*(1), 9–26.

Creswell, J. W. (1994). *Research Design: Qualitative & quantitative approaches*. Thousand Oaks, Calif.: Sage Publications.

Currie, W. L., & Swanson, E. B. (2009). Special issue on institutional theory in information systems research: Contextualizing the IT artifact. *Journal of Information Technology*, 24(4), 283–5.

Delanty, G. (ed.). (2006). *Handbook of Contemporary European Social Theory*. London, UK: Routledge.

DeSanctis, G., & Poole, M. S. (1994). Capturing the complexity in advanced technology use: adaptive structuration theory. *Organization Science, 5*(2), 121–47.

Di Maggio P.J. & Powell W.W. (1983) The Iron Cage Revisited: Institutional Isomorphism and Collective Rationality in Organizational Fields. *American Sociological Review* 48(2), 147–60.

Driedonks, C. F., Gregor, S., & Wassenaar, A. (2003). Economic and social analysis of the adoption of B2B electronic marketplaces: a case study in the Australian beef industry. Paper presented at the Sixteenth Bled eCommerce Conference: eTransformation, Bled, Slovenia, June.

Eisenhardt, K. M. (1989). Building theories from case study research. *Academy of Management Review, 14*(4), 532–50.

Foucault, M. (1977). *Discipline and Punish*. London, UK: Tavistock.

Friedman, T. L. (2006). *The World is Flat*. London, UK: Penguin Books.

Gibson, C. F., & Nolan, R. L. (1974). Managing the four stages of EDP growth. *Harvard Business Review, 52*(1), 76–87.

Glaser, B. G. (1978). *Theoretical Sensitivity: Advances in the methodology of grounded theory*. Mill Valley, Calif.: Sociology Press.

Glaser, B. G. (1998). *Doing Grounded Theory: Issues and discussions*. Mill Valley, Calif.: Sociology Press.

Glaser, B. G., & Strauss, A. L. (1967). *The Discovery of Grounded Theory: Strategies for qualitative research*. New York, NY: Aldine Publishing Company.

Gregor, S., & Jones, K. (1999). Beef producers online: diffusion theory applied. *Information Technology & People, 12*(1), 71–85.

Guba, E. G., & Lincoln, Y. S. (1994). Competing paradigms in qualitative research. In N. K. Denzin & Y. S. Lincoln (eds), *Handbook of Qualitative Research* (pp. 105–17). Thousand Oaks, Calif.: Sage Publications.

Kambil, A., & van Heck, E. (1998). Reengineering the Dutch flower auctions: a framework for analyzing exchange organizations. *Information Systems Research, 9*(1), 1–19.

Kaylor, C. H., Deshazo, R., & Van Eck, D. (2001). Gauging e-government: a report on implementing services among American cities. *Government Information Quarterly, 18*(4), 293–307.

Klein, A., & Krcmar, H. (2006). DCXNET: e-transformation at DaimlerChrysler. *Journal of Information Technology, 21*(1), 52–65.

Latour, B. (1993). *We Have Never Been Modern*. Cambridge, Mass.: Harvard University Press.

Lindsey, D., Cheney, P. H., Kasper, G. M., & Ives, B. (1990). TELCOT: an application of information technology for competitive advantage in the cotton industry. *MIS Quarterly, 14*(4), 347–57.

Malone, T., & Laubacher, R. (1998). The dawn of the e-lance economy. *Harvard Business Review, 76*(5), 144–52.

Markus, M. L. (1983). Power, politics and MIS implementation. *Communications of the ACM, 26*(6), 430–44.

Markus, M. L., & Robey, D. (1988). Information technology and organizational change: causal structure in theory and research. *Management Science, 34*(5), 583–98.

Misa, T. J. (2004). *Leonardo to the Internet: Technology and culture from the Renaissance to the present*. Baltimore, Md: Johns Hopkins University Press.

Orlikowski, W. J. (1993). CASE tools as organizational change: investigating incremental and radical changes in systems development. *MIS Quarterly, 17*(3), 309–40.

Orlikowski, W. J. (2000). Using technology and constituting structures: a practice lens for studying technology in organizations. *Organization Science: A Journal of the Institute of Management Sciences, 11*(4), 404–28.

Ouchi, W. G. (1980). Markets, bureaucracies, and clans. *Administrative Science Quarterly, 25*(1), 129–41.

Ridley-Duff, R. (2009). Cooperative social enterprises: company rules, access to finance and management practice. *Social Enterprise Journal, 5*(1), 50–68.

Rogers, E. M. (1995). *Diffusion of Innovation*. [Fourth edn]. New York, NY: The Free Press.

Rosenau, P. (1992). *Post-Modernism and the Social Sciences: Insights, inroads, and intrusion*. Princeton, NJ: Princeton University Press.

Rossignoli, C., Carugati, A., & Mola, L. (2009). The strategic mediator: a paradoxical role for a collaborative e-marketplace. *Electronic Markets—The International Journal on Networked Business, 19*(1), pp. 55–66.

Schön, D. A. (1983). *The Reflective Practitioner: How professionals think in action.* New York, NY: Basic Books.

Schön, D. A. (1987). *Educating the Reflective Practitioner*. San Francisco, Calif.: Jossey-Bass.

Scott Morton, M. S. (1971). *Management Decision Systems: Computer-based support for decision making*. Boston, Mass.: Division of Research, Graduate School of Business Administration, Harvard University.

Scott, W. R. (2008). *Institutions and Organizations: Ideas and interests*. Thousands Oaks, Calif.: Sage.

Senyucel, Z. (2008). Impact of ICTs on user–provider relations: perspectives from UK local authorities. *Information Technology & People, 21*(4), 401–14.

Sewell, G. (1998). The discipline of teams: the control of team-based industrial work through electronic and peer surveillance. *Administrative Science Quarterly, 43*(2), 397–428.

Spencer, H., & Carneiro, R. L. (1974). *The Evolution of Society: Selections from Herbert Spencer's Principles of sociology*. Chicago, Ill.: University of Chicago Press.

Stalk, G., Evans, P., & Shulman, L. (1992). Competing on capabilities: the new rules of corporate strategy. *Harvard Business Review, 70*(2), 57–69.

Svejvig, P., Carugati, A., (2010) Enterprise Systems are Hard to Die: On the Complexities of Decomissioning Enterprise Systems. Paper presented at the 15th Conference of The Association Information et Management (AIM 2010), La Rochelle, France.

Thompson, W. (1824). *An Inquiry into the Principles of the Distribution of Wealth Most Conducive to Human Happiness; Applied to the newly proposed system of voluntary equality of wealth*. London, UK: Longman, Hurst, Rees, Orme, Brown & Green.

Turner, J. H. (2000). Herbert Spencer. In G. Ritzer (ed.), *The Blackwell Companion to Major Social Theorists* (pp. 81–104). Malden, Mass.: Blackwell.

Tushman M.L. & Anderson P. (1986) Technological Discontinuities and Organizational Environments. *Administrative Science Quarterly* 31(3) 439–65.

Van de Ven, A. H., & Poole, M. S. (1989). Methods for studying innovation processes. In A. H. Van de Ven, H. L. Angle & M. S. Poole (eds), *Research on the Management of Innovation: The Minnesota studies* (pp. 31–54). New York, NY: Harper & Row.

Vitharana, P. & Dharwadkar, R. (2007), Information systems outsourcing: linking transaction cost and institutional theories, *Communications of the AIS*, 20, 346-370

Walsham, G. (1995). Interpretive case studies in IS research: nature and method. *European Journal of Information Systems, 4*(2), 74–81.

Zuboff, S. (1988). *In the Age of the Smart Machine*. New York, NY: Basic Books.

10. Competing with Business Analytics: Research in progress

Harjot Singh Dod
University of Melbourne

Rajeev Sharma
University of Melbourne

Abstract

Gaining competitive advantage is the focus of any organisation operating in today's marketplace. With readily available integrated data from data warehouses and innovation in business intelligence tools, managers are well placed to make smart decisions and hence better the competitive advantage of their firms. Only a few organisations have been able to fully realise the potential of business analytics (BA). A part of the reason could be a lack of understanding as to how business analytics can provide competitive advantage. A brief literature review on the topic of business analytics, which tries to capture the relevant research in this field, is presented. Literature on theory building is also analysed in order to provide a secure basis for the intended future research.

Introduction

Business analytics (BA) is a new and upcoming field in information systems (IS) research. Kohavi et al. (2002) argue that the strategic value of business analytics has led to significant development of business applications in areas that analyse customer data. These applications have been used to 'reduce customer attrition, improve customer profitability, increase the value of e-commerce purchases, and increase the response of direct mail and email marketing campaigns' (Kohavi et al., 2002, p. 47). Applications in other areas like finance, marketing, production, manufacturing, human resources and research and development have also been described in the literature (Davenport, 2006; Kohavi et al., 2002; Sharma et al., 2010). It has been argued that organisations will be able to create competitive advantage with the use of business analytics (Davenport, 2006). These claims are yet to be explained theoretically and researchers are trying to build theoretical models for the same (Sharma et al., 2010). The aim of this chapter is to review recommendations from the existing theory-building literature and analyse their significance for future theory-building initiatives to explain BA and its effect on competitive advantage.

The concept of business analytics is new in the IS literature. In order to better define this concept, we first discuss business analytics and its evolution. Since our aim is to build a theory analysing the competitive advantage provided by business analytics, we follow this with a discussion of the concept of competitive advantage. We then analyse the topic of theory building by presenting a brief literature review, and several recommendations from the literature are reviewed for the impact they might have on theory building for business analytics. Finally, we briefly analyse theories from relevant parent disciplines and the contribution they might make to theory building in business analytics.

Business Analytics

Business analytics involves acquiring new knowledge through an analysis of data and information from various sources, and employing that knowledge to develop and implement value-creating competitive actions (Sharma et al., 2010). The data analysed often reside in integrated databases and data warehouses and the analysis is often conducted employing tools such as data mining, visualisation, online analytical processing, statistical and quantitative analysis, and explanatory and predictive models.

Recent IS research in business analytics suggests that there is a statistically significant relationship between analytical capabilities and supply chain performance (Trkman et al., 2010). From a supply chain management perspective, business analytics is defined as 'a group of approaches, organisational procedures and tools used in combination with one another to gain information, analyse that information, and predict outcomes of problem solutions in any of the four areas (Plan, Source, Make, Deliver)' (Trkman et al., 2010, p. 318). Improving supply chain performance involves complex management processes such as identifying measures, defining targets, planning, communication, monitoring, reporting and feedback. This leads to the conclusion that making profitable business decisions based on very large volumes of internal and external data is only possible with business analytics that enables the analysis of data gathered in vast quantities on a regular basis (Trkman et al., 2010).

Evolution of Business Analytics

Using available information to aid decision making is natural for human beings. To understand how businesses are developing this capability, March and

Hevner (2007) use general systems theory to first understand what a business is from an economic point of view. They describe businesses as systems for transforming lower-valued resources acquired from the environment into higher-valued goods and services for the environment. With profit making as the primary motive, it becomes important that organisations systematically acquire relevant information to make better decisions and stay ahead of their peers. Data warehouses are playing the role of information processors by integrating, interpreting and transforming the data for managerial decision making (March and Hevner, 2007). A data warehouse can be metaphorically thought of as a facilitating manager, helping to better understand the business problems, opportunities and performance avenues.

Another IS innovation linked with gathering data is enterprise resource planning (ERP). The widespread implementation of ERP systems helps organisations to integrate their information and make the integrated information readily available for further use. The implementation of ERP systems along with the improvement and adoption of data warehousing technology can help decision support systems to access customised data. But there are two important issues that need to be addressed before ERP systems can be used in decision support systems. The first is that ERP systems are built from a transaction processing point of view and hence can provide only ad-hoc decision support at best (Seethamraju, 2007). To cover this gap, organisations are increasingly bolting a decision support system on top of their existing ERP systems (Seethamraju, 2007). The second issue concerns external information, which is often very relevant in aiding strategic decisions and is typically not captured in an ERP environment, thus limiting the use of ERP for operational decision support.

With advances in online analytical processing (OLAP) and data-mining tools, data warehouses are becoming increasingly efficient in analysing and reporting. Analysing structured data has become relatively easy. There are numerous other information sources that are unstructured or at best semi-structured (email, web pages, research papers, reports, and so on). Often overlooked and under-utilised, these sources are nevertheless important in ensuring the quality of decision making and, as finding, organising and analysing these sources is difficult, mastering this is quite likely to give greater competitive advantage to a firm that does so. Recently IS researchers have focused on assimilating semi-structured and unstructured data with the structured data to provide a comprehensive view for strategic decision making, with Baars and Kemper (2008) prescribing three approaches in this area.

The three approaches identified by Baars and Kemper (2008) vary from a simple integrated presentation of different types of data sources to applying complex knowledge management techniques to diffuse relevant data across the organisation. The first approach is called 'integrated presentation', where both structured and unstructured data can be simultaneously accessed via an integrated user interface. The second approach is called 'analysis of content collection', where it is possible to analyse unstructured data based on its metadata (date of creation, length, author, and so on). The third approach is called 'distribution of analysis results and analysis templates', where it is presupposed that there is business intelligence knowledge that is of relevance to users and that can be efficiently shared. Moreover, even if the concrete results are too specific for immediate reuse, the process of deriving those results (analysis model used, selected data sources, visualisations used, and so on) might be useful for other users. The choice of a particular approach is based on the relative business potential of integrating varied data sources using the different approaches (Baars and Kemper, 2008).

Apart from the technical issues, IS research has also focused on the problems faced by managers after implementation of a business intelligence or business analytics system. A technically successful implementation does not necessarily lead to desired benefits. Several practical issues have to be monitored when organisations are using the system. New information needs, change in organisational focus and unexpected market developments among other factors are the realities of a dynamic business environment (Dekkers et al., 2007). A greater focus on organising a business intelligence/business analytics system initiative to answer the dynamic information needs of business users is required.

Competitive Advantage

Among other things, the development of business analytics as a tool is enmeshed in an organisation's use of other new and innovative information technology (IT) tools. Therefore, before researching the role of business analytics in creating a competitive advantage, it would be beneficial to understand how IT has been traditionally employed in trying to build competitive advantage. It can be argued that investment in information technology helps to facilitate and enhance the work done by an organisation but not all IT products developed by an organisation are useful, and in some cases they can leave a deep hole in the pockets of investing companies. Carr (2004) goes to one extreme by arguing that IT, which requires a lot of investment and often does not provide any competitive advantage, has become a commoditised entity in today's corporate world.

IT and Competitive Advantage

With many IT projects failing to deliver their promised results, IS researchers have focused their energy towards finding the causes. The first consensus reached was that stand-alone investment in IT will not necessarily lead to any addition of economic value. Information technology networks and databases can be easily procured from the technology market and hence are unlikely to be the source of a distinct competitive advantage. Much focus is required on 'the configuration of an activity system, dependent on IT at its core, which fosters the creation and appropriation of economic value' (Piccoli and Ives, 2005, p. 748). But even with the effective implementation of IT, organisations are now concerned about the easy imitability of their IT initiatives as new and improved off-the-shelf products are made available to their competitors. Quick obsolescence is another problem fuelled by the fast-paced innovation in IT offerings. In this dynamic world of IT, maintaining a competitive advantage is becoming increasingly difficult.

There is a plethora of literature that talks about how organisations can use IT to create competitive advantage. Piccoli and Ives (2005, p. 751) take the resource-based view to identify four barriers to erosion of competitive advantage that might be established. These barriers are the 'IT resources barrier, complementary resources barrier, IT project barrier and pre-emption barrier'. Ravichandran and Lertwongsatien (2005) measure firm performance by analysing how IT is used effectively to support and enhance the firm's core competencies. They recognise that firm resources are the main drivers of firm performance and suggest using IT to reinforce firm resources, thus creating complementary resources. Bhatt and Grover (2005, p. 253) discuss whether IT can be used to provide differential benefits to individual firms. They frame 'value, competitive and dynamic capabilities' as three distinct types of capabilities and describe the relationship between these capabilities and competitive advantage.

Business Analytics and Competitive Advantage

While the research on IT and competitive advantage is still evolving, a parallel discussion on gaining competitive advantage through the use of the vast amounts of data held by today's organisations has gained momentum. According to Davenport (2006), businesses today are offering similar types of products using comparable technologies, which renders competing on business processes alone unviable. He provides various instances where business analytics is being used in a variety of sectors like entertainment (Harrah's), finance (Capital One), e-commerce (Amazon), logistics (UPS), consumer products (P&G) and retail (Wal-Mart). He further argues that business analytics has helped these companies to intelligently use their existing data sources so as to aid decision making and build a remarkable competitive advantage.

A common claim across sectors is that the use of business analytics has a direct positive impact on competitive advantage. Analysing such intelligent use of data sources is a complex and evolving research stream. Given the need of businesses to effectively utilise their data, future research in IS should focus on explaining how companies can best use their data to build competitive advantage.

Since business analytics is a relatively new term, a lot of scope exists for conceptualising this area and theorising about how it can provide competitive advantage for organisations, so the next section presents some suggestions from the IS literature on theory building and how these can be used in our future endeavours to theorise about the competitive advantage provided by business analytics.

Theory Building for Business Analytics

Given the substantial recent focus on publishing 'good theory' papers, researchers nowadays are highly motivated to build a theory in their field. But, in order to build a strong theory, researchers need to take care of several aspects, which include: presenting logical arguments explaining the empirical relationships found in the data, explaining the strong predictive capabilities of certain variables, explaining the proposed causal or other connections (usually in a diagram) and explaining the rationale for a hypothesis. A strong theory 'delves into underlying processes so as to understand the systematic reasons for a particular occurrence or non-occurrence' (Sutton and Staw, 1995, p. 378).

One such theory-building initiative in business analytics was the development of a dynamic business analytics capabilities (DBAC) framework, which aims to explain the competitive advantage provided by business analytics (Sharma et al., 2010). The DBAC model focuses on how knowledge generated by a business analytics infrastructure can be used to create organisational factors and processes that in turn lead to the development and implementation of value-creating competitive actions. Figure 10.1[1] shows a representation of the DBAC model as conceptualised by Sharma et al. (2010).

A meta-theoretical analysis by Gregor (2006) presents and defines multiple views of theory in information systems. It furthers the argument that 'theories are practical' because of the systemic manner in which knowledge is accumulated and is made available for professional practice. The 'systematic manner of knowledge accumulation' is described by identifying the nature of five types of IS theory (analysis, explanation, prediction, explanation and prediction, and design and action). Accordingly, a good theory in business analytics should

---

1 Figure used with permission from the original authors.

strive to present knowledge about business analytics in a 'systematic manner' through the use of various factors (variables, constructs, concepts, and so on) whatever position it occupies in Gregor's typology.

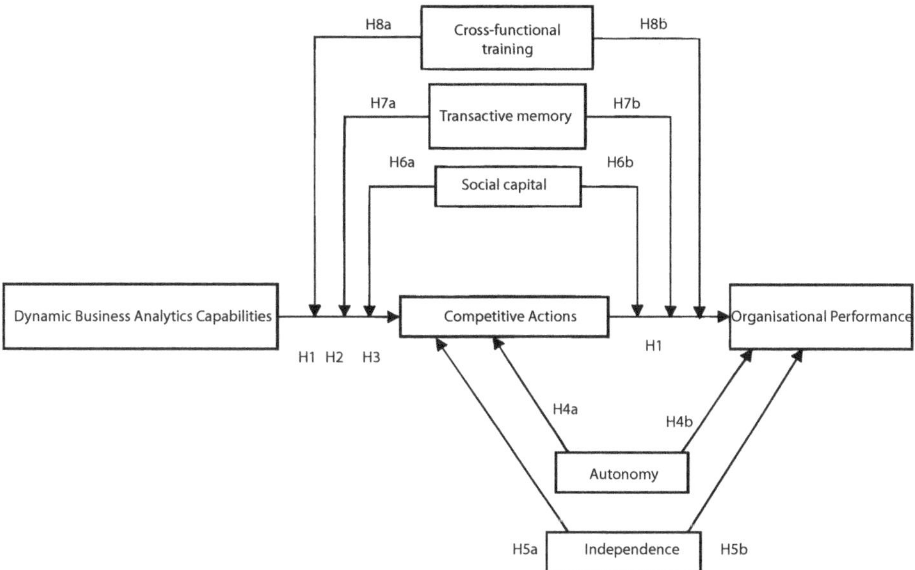

Figure 10.1 The Dynamic Business Analytics Capabilities (DBAC) Model

In order to achieve a 'complete' theory, as defined by Whetten (1989), it is necessary to take care of four essential elements. We now analyse the DBAC theory presented by Sharma et al. (2010) with respect to these four essential elements. First, we have to identify certain factors (variables, constructs, concepts) that would be the key elements for building a theory in business analytics (the 'what' element). From a technological perspective, these factors could be data warehousing, business intelligence, decision support and other systems. In the DBAC model, several organisational factors like cross-functional training, transactive memory, social capital, autonomy and independence are identified, which in turn are linked to a dependent variable called organisational performance.

Second, we need to describe the relationships among the identified factors (the 'how' element). Such relationships can be based on observations from industry, anecdotes and other forms of cognition. In the DBAC model, relationships between the factors are proposed in the form of several logically derived hypotheses (Sharma et al., 2010). Third, it is necessary to justify the selection of these few factors and the relationships between them (the 'why' element). This involves explaining the choice of factors and causal relationships logically to peers, for further scrutiny. The DBAC model argues that dynamic capabilities are independent of IT investment (Sharma et al., 2010). This assumption helps the authors to justify their focus on organisational factors. Further, the authors provide logical arguments for the selection of organisational parameters that might affect the firm's competitive advantage. Fourth, setting boundaries about the generalisability of the theory helps to justify the scope of theory (the 'who, where, when' element). Generalisability becomes an important topic when we consider the practical applicability of a theory.

In their business analytics research, Sharma et al. (2010) use deductive logic to construct their DBAC model. They draw on the literature on *dynamic capabilities* and define DBAC as a specific dynamic capability that utilises the operational and other data available in an organisation's information assets to develop resources and implement value-creating competitive actions (Sharma et al., 2010). Although the DBAC model identifies the key factors necessary to achieve performance gains from business analytics, there still is a need to empirically test the framework in industry settings. The scope of the DBAC model would be set out in the empirical testing phase where the testers would define the testing environment. This testing environment could then shape the generalisability of the DBAC model to other environments.

Evidence of the use of business analytics is present in multiple industries (Davenport, 2006). The ways to approach theory building in business analytics also vary, from the use of the dynamic capabilities literature in the DBAC model (Sharma et al., 2010) to the use of the supply chain management literature (Trkman et al., 2010). This implies that theories developed for business analytics might have to cater to a wide variety of industries, making the research more interdisciplinary. Researchers will have to be careful when generalising business analytics theories from one topic area to another.

A type of descriptive analysis can be undertaken to understand the practice of business analytics in industry. This effort can be classified as theory for analysis (Gregor, 2006), and empirical techniques like surveys and interviews can be used for exploration. At the same time, there is a need for qualitative analysis in this field, which among other things would talk about the perceived impact of business analytics in an organisation.

Theoretical Orientation from Parent Fields

It is evident that the development of information systems in areas like data mining, decision support systems, data warehousing and business intelligence has collectively shaped the business analytics arena (Davenport, 2006). Given this, well-established theories in these parent fields are likely to influence theory building in business analytics. One example is the 'multi-layer framework for business intelligence'.

The multi-layer framework for business intelligence conceptualised by Baars and Kemper (2008) is an important business intelligence framework that maps the logical business intelligence (BI) components and their core interrelations. It is important for future theory-building efforts in business analytics because this framework suggests how data can be collected from various sources, stored, analysed and distributed to relevant users. A business analytics researcher could take the concept further by theorising about how, after receiving the required information, users could employ it to gain competitive advantage in the marketplace. The framework was developed over the course of several years in tight interaction with practitioners from both the supply and the application sides (Baars and Kemper, 2008).

The framework distinguishes three layers: data, logic and access. The data layer is responsible for storing structured and unstructured data for management support functions. Usually, structured data are kept in special data repositories—data warehouses, data marts and operational data stores—while unstructured content is handled with content and document management systems. The data are extracted from source systems, which might include operational systems like ERP or SCM systems and external data sources (Baars and Kemper, 2008).

The logic layer provides functionality to analyse structured data or unstructured content and supports the distribution of relevant knowledge. The analytical functionality of the logic layer not only includes OLAP and data mining but also functionality to generate (interactive) business reports and perform ad-hoc analysis (Baars and Kemper, 2008). The access layer allows a user to conveniently access all relevant functions of the logic layer in an integrated fashion within the confines of defined user roles and user rights. Usually the access layer is realised with some sort of portal software that provides a harmonised graphical user interface (Baars and Kemper, 2008).

Conclusion

We argue that business analytics as a construct has not been properly defined in the literature yet so we have tried to provide some insight in this area. Nevertheless, further work needs to be done in order to analyse how various parent fields can contribute to the development of business analytics. Various aspects of theory building have also been reviewed and the role this literature could play in guiding theory-building initiatives for business analytics is also briefly analysed. Our aim is not to prescribe any particular method but rather to make researchers aware of what is recommended in the existing literature and how it can be used for research in business analytics.

References

Baars, H., & Kemper, H.-G. (2008). Management support with structured and unstructured data—an integrated business intelligence framework. *Information Systems Management, 25*(2), 132–48.

Bhatt, G. D., & Grover, V. (2005). Types of information technology capabilities and their role in competitive advantage: an empirical study. *Journal of Management Information Systems, 22*(2), 253–77.

Carr, N. G. (2004). IT doesn't matter. *IEEE Engineering Management Review, 32*(1), 24–32.

Davenport, T. H. (2006). Competing on analytics. *Harvard Business Review, 84*(1), 99–107.

Dekkers, J., Versendaal, J., & Batenburg, R. (2007). Organising for business intelligence: a framework for aligning the use and development of information. *Proceedings of the 20th Bled eConference eMergence, Bled, Slovenia*, 625–36.

Gregor, S. (2006). The nature of theory in information systems. *Management Information Systems Quarterly, 30*(3), 611–42.

Kohavi, R., Rothleder, N. J., & Simoudis, E. (2002). Emerging trends in business analytics. *Communications of the ACM, 45*(8), 45–8.

March, S. T., & Hevner, A. R. (2007). Integrated decision support systems: a data warehousing perspective. *Decision Support Systems, 43*(3), 1031–43.

Piccoli, G., & Ives, B. (2005). Review: IT-dependent strategic initiatives and sustained competitive advantage: a review and synthesis of the literature. *MIS Quarterly, 29*(4), 747–76.

Ravichandran, T., & Lertwongsatien, C. (2005). Effect of information systems resources and capabilities on firm performance: a resource-based perspective. *Journal of Management Information Systems, 21*(4), 237–76.

Seethamraju, R. (2007). ERP systems and decision support—an exploratory study. *Proceedings of the International Conference on Decision Support Systems (!CDSS 2007)*.

Sharma, R., Reynolds, P., Scheepers, R., Seddon, P., & Shanks, G. (2010). Business analytics and competitive advantage: a review and a research agenda. Paper presented at the Fifth International Federation of Information Processing WG 8.3 International Conference on Decision Support Systems, Lisbon, Portugal.

Sutton, R., & Staw, B. (1995). What theory is not. *Administrative Science Quarterly, 40*(3), 371–84.

Trkman, P., McCormack, K., de Oliveira, M. P. V., & Ladeira, M. B. (2010). The impact of business analytics on supply chain performance. *Decision Support Systems, 49*(3), 318–27.

Whetten, D. A. (1989). What constitutes a theoretical contribution? *Academy of Management Review, 14*(4), 490–5.

Part Three: The Big Picture

11. Theory: An informatics perspective

Craig McDonald
University of Canberra

Abstract

This chapter explores the concept of theory from an informatics perspective. That is, it frames a theory as a conceptual pattern, an information construct or ontology, and theory building as a process that creates uses and modifies a pattern. An agenda for information systems as a discipline for explicit pattern management is proposed.

Introduction

Suppose you are driving your car out in the country on a fine day. You are listening to a radio play on the stereo, watching the road, enjoying the view and the whole motoring experience. A police car appears out of nowhere and invites you to pull over. You do so, get out of your car and talk with the policeman. He says you were driving over the speed limit, that the speed limit on that stretch of road is 100 km/h and you were driving at 111 km/h according to his radar. In accordance with the *Motor Traffic Act*, the policeman says, he will issue you with a speeding fine. You object, saying that you were not travelling at the claimed speed. You ask your car computer to print out your data for the past 10 minutes, which it does, and the printout shows that 105 km/h was your fastest speed in that time.

There are many ways of framing this simple scenario. A sociologist might frame the scenario in terms of the power relations between driver and policeman; a psychologist might frame it in terms of mental states and disposition; a technologist might focus on the differences between the speed-recording devices; a police commander in terms of the fine income, and so on. The frame one selects will determine the kinds of concepts and relationships between concepts that are brought to bear on and extracted from the scenario. In what follows, we ask how this scenario could be seen in an informatics frame and derive, from what we find, some implications for theory and theory building in the information systems (IS) discipline.

A 'Grand Theory'

We can interpret the simple tale of the previous section using an old 'grand theory' that captures the essence of the informatics concept. This theory argues that there are three different kinds of things in the universe; that there are three different 'worlds', as Popper (1972) calls them. Things of the first kind are physical objects (roads, cars, speed-limit signs, and so on), apparent forces (gravity, wind, and so on) and their interactions. The disciplines that study this physical 'world one' include physics, chemistry, ecology, astronomy, and so on. Things of the second kind are cognitive processes—thinking, consciousness and creativity in the mind (including intelligence of many forms, both natural and artificial)—that are studied by psychology, artificial intelligence, linguistics, education, and so on. Third comes the semantic world of information comprising data, statements and articulated knowledge that are carried in language and that are studied by the informatics disciplines including information science, information systems, statistics, library studies, mathematics, communication, journalism, and so on. One form of interdisciplinarity involves studying the interactions between the worlds and this includes such disciplines as communications, social sciences, economics, and so on.

These three worlds interact: minds perceive the physical world, they interpret what they perceive in terms of their beliefs, values and their understanding of the social situation, they express their thoughts in statements and in bodily actions. In our scenario there are two minds: yours and the policeman's; both have perceptions of your car's speed, the speed limit at the relevant place, and both have understandings of the *Motor Traffic Act* (the road rules) although these perceptions and understandings might be quite different. You communicate through statements in a common language but nevertheless probably use that language sufficiently differently so that communication is less than perfect. Your car's computer and the policeman's radar have artificial intelligence—that is, they have 'perceptions' of the physical world and they 'express' their 'understanding' in 'language'. For a longer discussion of the three-worlds idea, see McDonald (2002).

Figure 11.1 shows a diagram of the scenario seen through the three-worlds conceptual lens.

This way of seeing the universe, this old 'grand theory', is not new; versions of it date back at least to Plato's cave allegory and to Aristotle. The elaboration by Karl Popper is, though, perhaps the best known of the more recent versions. In Popper's view, reality is divided into three parts: 'first, the world of physical objects or of physical states; secondly the world of states of consciousness, or of mental states, or perhaps of behavioural dispositions to act; and thirdly, the world of objective contents of thought' (Popper, 1972, p. 106).

Popper distinguished thought, in the sense of the content of statements, and thought, in the sense of thought processes, as belonging to two entirely different worlds. 'If we call the world of "things"—of physical objects—the first world, and the world of subjective experiences (such as thought processes) the second world, we may call the world of statements in themselves the third world' (Popper, 1976, pp. 180–1).

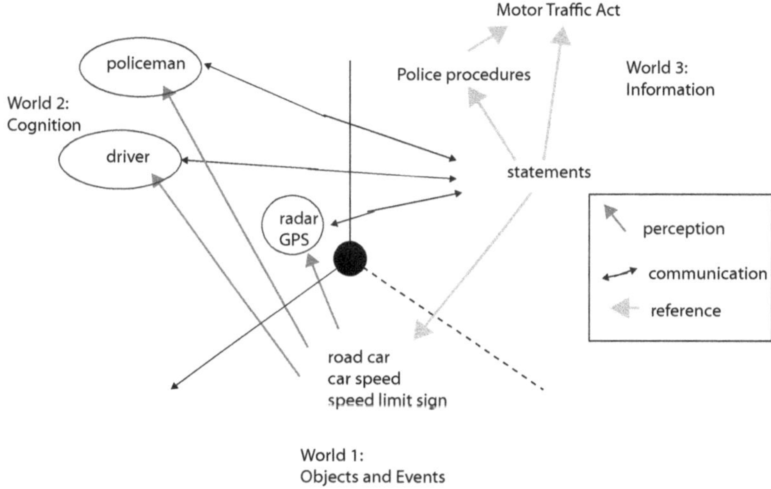

Figure 11.1 Popper's Three-World Model Illustrated Using the 'Speeding' Scenario

The 'grand theory' accommodates other ways of understanding informatics as, for example, the 'meaning triangle' (Odgen and Richards, 1923) shown in Figure 11.2. Similarly, Figure 11.3 shows how semiotics (after Peirce) and the three-worlds view fit together. There are also many other philosophies that rest on a three-part view of the world: the physical, the mental and the abstract.

The scenario described at the start of this chapter involved two minds, both formed through quite different generic and experiential histories and so seeing the world quite differently when the scenario begins. If the 'information' each received was the same, they would interpret it differently, but of course the 'information' each receives is not the same. So the interaction between these two minds is partly about vehicle speed in world one, partly about their own world-two states, and partly about the *Motor Traffic Act* in world three.

Of course a mind perceives and interprets masses of signals. A grand theory, alternative to this one, might start from the mind and conceive of 'information' as being anything that informs the mind ('environmental information', as Floridi, 2010, would have it). The purpose of the grand theory in this chapter, however, is to provide an account of the world for IS use, not for a study of mind.

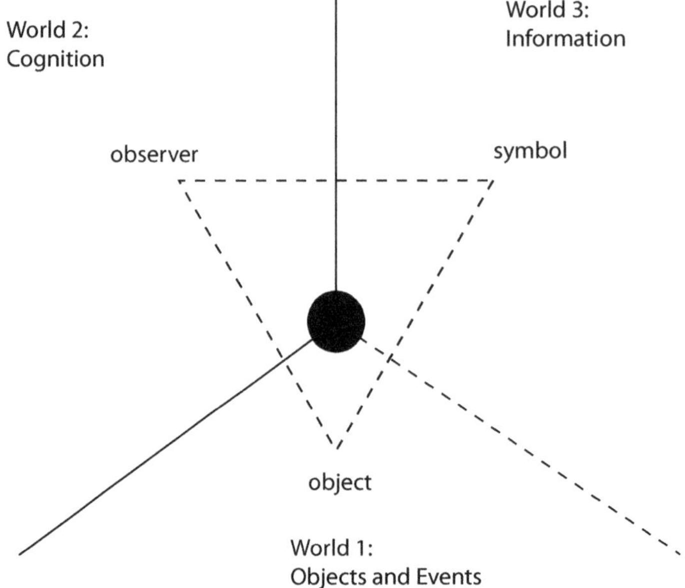

Figure 11.2 The Meaning Triangle

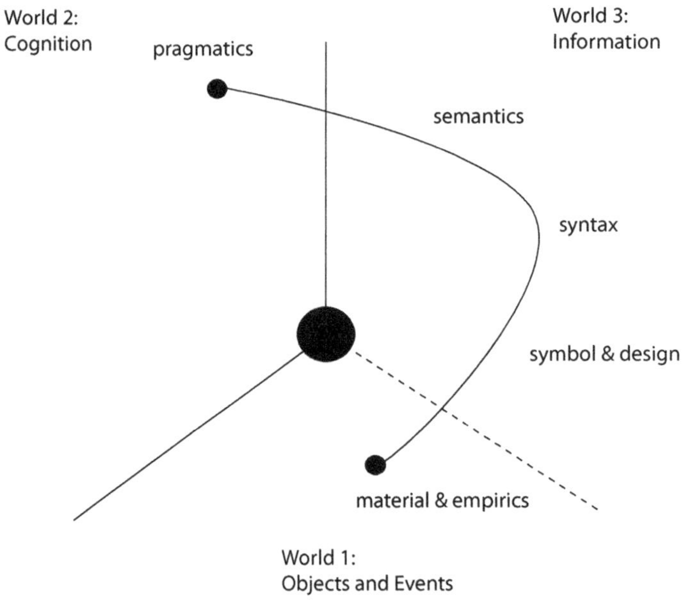

Figure 11.3 Semiotics and Popper's Three Worlds

One way to unpack world three, from an IS perspective, is to consider the different forms that information can take. The history of modern IS is, arguably, grounded in the development of electronic data-processing (EDP) systems and built on subsequent computing technologies. Electronic data-processing systems recognise two forms of information. The first is data that take the form of a statement asserting the value of an attribute of an object at a point in time. An example would be the temperature (attribute) of this room (object) being 21 degrees (value). The second form information can take is process knowledge describing how data can be processed. This articulated knowledge may be in the form of processes and procedures or coded into computer executable algorithms, scripts, and so on. The role of IS was to design systems that:

- represented an understanding of an information domain in terms of data and the knowledge of how to process them
- executed that understanding by capturing data and processing them effectively for human use.

Such systems are tightly governed by an ontology, represented in meta-data and rule specifications. While ontology-governed systems remain the backbone of the IS professional focus, communications systems dealing with other forms of digital objects usually operate in parallel. Communication systems, including email, social media, and so on, accept, transmit and display content that is not rigorously governed. From a technology perspective, the acronym ICT (information and communication technologies) is an indicator of separation between systems of information and those of communication.

Some Anomalies in the 'Grand Theory'

'Grand theories' are useful in all disciplines to contextualise and bring together more empirical theory work in the domain. They give a sense of wholeness to a body of work and provide the large ontological categories that allow researchers to better ground their work or at least to challenge more orthodox views. There is a tendency in the informatics disciplines, especially when humanities and social sciences approaches are brought in, to invent jargon and reuse concepts so as to make a 'new' research field incommensurable with the existing work. While this tendency might resonate with some universities, funding bodies and conferences, it fragments the serious descriptive, explanatory, predictive and design powers of the discipline.

All theory is tentative and subject to test and challenge. The grand theory is no exception. It might need some revision in the light of technical change as well as other inherent theoretical issues.

Some technology changes were indicated above in the discussion of information forms, and new phenomena that appear need to be analysed to see how they fit with existing high-level understandings. With this in mind, note that world-three statements are concepts and relationships, so images and signal datasets do not qualify as citizens of world three. They remain world-one entities until they are processed and some information is drawn from them and posted into world three.

World three is a very literal place. It comprises statements—where a statement is a structure of concepts or categories whose instances are referred to. Data statements have a single instance, knowledge statements refer to instances of a particular kind and text statements are rather a mixed bag. As these statements exist independently of their author, they are open to examination and test to determine their veracity; their truth or falsehood. But it is not always easy to tell whether statements are literal or not. Much human communication in language is metaphoric, poetic or intentionally misleading. For example, Frankfurt (2005) gives a description of insidious non-literal statements:

> It is impossible for someone to lie unless he thinks he knows the truth.
>
> Producing bullshit requires no such conviction.
>
> A person who lies is thereby responding to the truth, and he is to that extent respectful of it.
>
> When an honest man speaks, he says only what he believes to be true; and for the liar, it is correspondingly indispensable that he considers his statements to be false.
>
> For the bullshitter, however, all these bets are off: he is neither on the side of the true nor on the side of the false.
>
> His eye is not on the facts at all, as the eyes of the honest man and of the liar are, except insofar as they may be pertinent to his interest in getting away with what he says.
>
> He does not care whether the things he says describe reality correctly.
>
> He just picks them out, or makes them up, to suit his purpose. (Paraphrased from pp. 55–6)

Margaret Thatcher (1987) famously said '[t]here is no such thing as society'. This is puzzling. In the grand theory, information is statements that refer to something, so there needs to be 'society' for the statement to be literal, but if there is 'society' then the statement cannot be true. Perhaps this is one for the new subcategory 'statements not to be taken literally'. But it raises the question of what kind of concept is 'society'?

A similar question arises from the scenario at the start of this chapter. The *Motor Traffic Act* is a law that governs the behaviour of police and drivers alike. It is a set of statements, but has a very different role in world three than do data or text statements. It is a referent in its own right. Ontologies, vocabularies and theories are also concepts that have been mentioned in this chapter and they are all alike in being referred to by statements, but which do not do any referring themselves. They are instead long-term, stable patterns that govern discourse and behaviour and that have strong social systems that support and enforce them.

A 'Grand Theory' Modification

An important new part of world three needs to be created: patterns. Figure 11.4 shows the place of the pattern part of world three with some examples of significant types of patterns. The figure also gives an indication of their degree of specification formality and the degree of social deliberation that goes into their creation and maintenance.

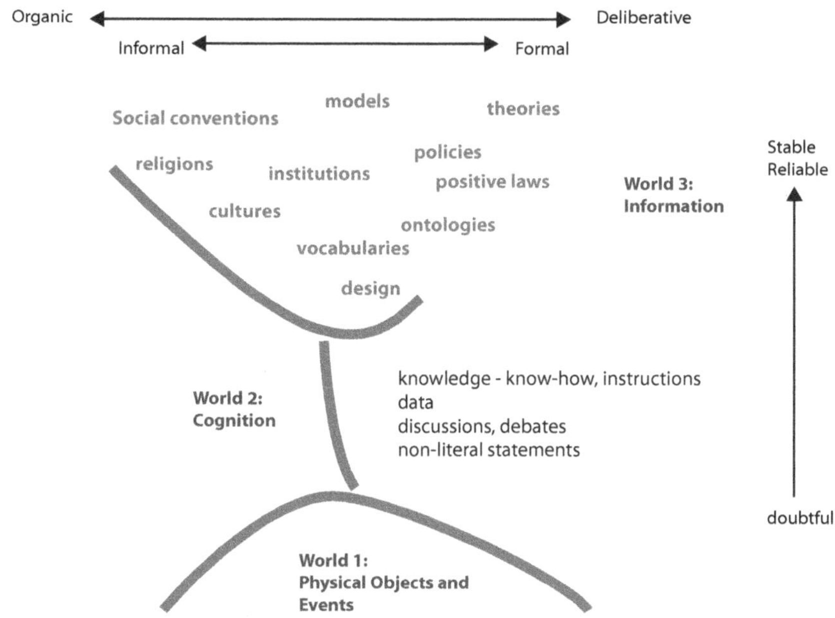

Figure 11.4 The Place of Patterns in World Three

The identification of a pattern subclass in the information world better accounts for the phenomena identified above as problematic. It is a place for articulated conceptual structures, for the big ideas that lurk in the background of our everyday lives. The grand theory does not suggest that patterns are based in physical reality; some of the most powerful patterns have little to do with the physical world. There are big questions about the conceptual structure of English society, but Thatcher was wrong to say that it does not exist since it is a pattern.

Social processes create and modify patterns that give them their 'warrant' with a population. Patterns are rarely the product of one mind.

Patterns are internalised in each world-two mind more or less faithfully through its experience, acculturation and education. Each mind has its own knowledge, and that knowledge changes over time. One's internalised patterns underlie perceptions and actions. External patterns are a touchstone for reflecting on why one thinks and acts the way one does.

Patterns provide the referent necessary to test statements, expose bullshit, and so on. Attempts to more formally articulate informal patterns might reveal just how little we really know about the non-physical world humanity has built.

Theory and Theory Building

Theory can be seen as an articulated pattern of concepts and relationships that has gained stable, reliable status in academic disciplines through a process of research, debate and publication and which functions to describe, explain and predict phenomena (Gregor, 2002). The conceptual graphs formalism of Sowa (1984, 2000) provides a means to represent precisely the components and structure of a theory. If a theory cannot be represented this way, perhaps it still contains connotation rather than being denotative.

In his discussion of theory and theory building, Weber (2003) concentrates on the constructs, laws, states and events that make up theory. Like so much theoretical writing, the position is sound, but fails to recognise its own theoretical context. It does not recognise any larger theoretical space of which it is a part, or in what way it is a part of that space. It does not recognise its similar but different siblings.

Much IS research is characterised by weak articulation of its theoretical context, so the discipline appears fragmented and has had difficulty building coherent, stable theory. The *Theories Used in IS Research Wiki* (BYU, 2011), for example, is a useful catalogue but it is an unstructured collection. The hard conceptual work of theorising about this collection of theories at the next level of abstraction seems still to be done.

Literature views in IS theory building are supposed to provide a conceptual framework for a research project. Unfortunately, they often seem to simply identify other work in the field of interest instead of producing a conceptual framework that locates the field upwards into a general theoretical context and sideways to distinguish it from adjacent or overlapping work. Typically, all that is done is to propose a concept and work downwards to operationalise variables for analysis. Information systems journal editors, research supervisors, examiners, reviewers and research educators would do well to heed Christopher Alexander's words:

> [W]hen you build a thing you cannot merely build that thing in isolation, but must also repair the world around it, and within it, so that the larger world at that one place becomes more coherent, and more whole; and the thing which you make takes its place in the web of nature, as you make it. (Alexander et al., 1977, p. xiii)

Conclusion

Perhaps one of the most longstanding interests of the informatics disciplines (those that study information) has been the representation of our understanding of the world and the construction of human and technical systems that deploy that understanding.

Grand theory is important for two reasons. First, it puts the informatics disciplines in context and is itself a pattern against which disciplinary debate can be conducted. Second, because patterns are an information construct, they are susceptible to examination and systems building by the IS discipline.

There is vast scope for IS research to be done on the specification of different pattern types, technology for representation and deployment, systems for theory building, quality of representation and lines of evidence, data curation and reuse, and industry deployment through instruments and designs. Information systems should be defining the human and systems aspects of the development and maintenance of the pattern part of world three.

Ontological technologies for industry, semantic web development and e-research are all developing technologically without the systems context that IS should bring to them. Information domains including legal informatics, health informatics and government informatics are all adopting ontologies without the systems aspects that IS should bring to them. Yet IS does not seem to see these trends or its place in them.

This chapter has suggested a modification to a grand theory of informatics to accommodate patterns of various kinds to better account for information phenomena. It has argued for IS theory and theory building to become more coherent and theoretically integrated using patterns and for IS to take on pattern explication as an area of research and development.

References

Alexander, C., Ishikawa, S., Silverstein, M., Jacobson, M., Fiksdahl-King, I., & Angel, S. (1977). *A Pattern Language: Towns, buildings, construction*. Oxford, UK: Oxford University Press.

BYU (2011). *Theories Used in IS Research Wiki*, <http://istheory.byu.edu/wiki/Main_Page>

Floridi, L. (2010). *Information: A very short introduction*. Oxford, UK: Oxford University Press.

Frankfurt, H. G. (2005). *On Bullshit*. Princeton, NJ: Princeton University Press.

Gregor, S. (2002). A theory of theories in information systems. In S. Gregor & D. Hart (eds), *Information Systems Foundations: Building the theoretical base* (pp. 1–10). Canberra, ACT: ANU E Press.

McDonald, C. (2002). Information systems foundations—Karl Popper's third world. *Australasian Journal of Information Systems, 10*(1), 59–69.

Odgen, C. K., & Richards, I. A. (1923). *The Meaning of Meaning*. New York, NY: Harcourt Brace.

Popper, K. R. (1972). *Objective Knowledge, An Evolutionary Approach*. Oxford, UK: Clarendon Press.

Popper, K. R. (1976). *Unended Quest: An intellectual autobiography*. New York, NY: Fontana.

Sowa, J. F. (1984). *Conceptual Structures: Information processing in mind and machine*. Boston, Mass.: Addison-Wesley.

Sowa, J. F. (2000). *Knowledge Representation: Logical, philosophical and computational foundations*. Pacific Grove: Calif.: Brooks/Cole.

Thatcher, M. (1987). <http://www.margaretthatcher.org/document/106689>

Weber, R. (2003). Theoretically speaking. *MIS Quarterly, 27*(3), iii–xii.